SpringerBriefs in Geography

SpringerBriefs in Geography presents concise summaries of cutting-edge research and practical applications across the fields of physical, environmental and human geography. It publishes compact refereed monographs under the editorial supervision of an international advisory board with the aim to publish 8 to 12 weeks after acceptance. Volumes are compact, 50 to 125 pages, with a clear focus. The series covers a range of content from professional to academic such as: timely reports of state-of-the art analytical techniques, bridges between new research results, snapshots of hot and/or emerging topics, elaborated thesis, literature reviews, and in-depth case studies.

The scope of the series spans the entire field of geography, with a view to significantly advance research. The character of the series is international and multidisciplinary and will include research areas such as: GIS/cartography, remote sensing, geographical education, geospatial analysis, techniques and modeling, landscape/regional and urban planning, economic geography, housing and the built environment, and quantitative geography. Volumes in this series may analyze past, present and/or future trends, as well as their determinants and consequences. Both solicited and unsolicited manuscripts are considered for publication in this series.

SpringerBriefs in Geography will be of interest to a wide range of individuals with interests in physical, environmental and human geography as well as for researchers from allied disciplines.

Camilo Vladimir De Lima Amaral

The Urban Reproduction of Subjectivities

Deconstructing Neoliberal Architectures in London

Springer

Camilo Vladimir De Lima Amaral
Interuniversity Department of Regional and Urban Studies and Planning
Politecnico di Torino
Turin, Italy

ISSN 2211-4165 ISSN 2211-4173 (electronic)
SpringerBriefs in Geography
ISBN 978-3-032-01464-1 ISBN 978-3-032-01465-8 (eBook)
https://doi.org/10.1007/978-3-032-01465-8

This Springer imprint is published by the registered company Springer Nature Switzerland AG
The registered company address is: Gewerbestrasse 11, 6330 Cham, Switzerland

Preface

In general, common sense would accept the idea that 'architecture has always shaped society' (Rebel Architecture 2014). Monuments, squares, streets, patterns, houses, nature, and rituals, among others, are understood as elements that shape cultures and ways of life in every city. By contrast, we increasingly accept the idea that architecture and urban design cannot change society because it is a small part of a complex socioeconomic structure. This fundamental paradox lies in design's simultaneous ability to shape society while being unable to reshape it.

Initially, this book maps how the current debate in architecture frames this paradox through a series of deadlocks. These 'frames' may reveal how the debate creates false dilemmas and arguments for moderation, which result from internal disputes that divert radical inquiries about the social role of spatial design practices.

The paradox (to shape and yet be unable to shape) obscures a fundamental contradiction in the field's epistemology. Normally, architecture is theorized as a 'given' rather than a social product. Yet, the architecture of the city (to use Aldo Rossi's expression) is not a natural fact; it is a social result of human activity in a given social, historical and economic context. Moreover, architecture is not merely a random, free or spontaneous activity. It involves a complex system of knowledge and skills, passed down through generations. In this sense, history is the foundation of the architecture of the city.

Furthermore, the practice of shaping the city is also framed by disciplinary forces. As a 'profession', a field of 'knowledge', a system of 'corporatist' interests, a shared cultural language, and a social praxis, a set of skills or a division of labour, it regulates patterns of behaviour, defines measures of comparison, controls interactions, imposes ways of thinking, and determines what is seen and unseen. It also establishes social hierarchies, reinforces identities and roles, and structures everyday life. At the very least, architecture and urban design are currently designed to function this way and reproduce this form of subjectivity.

This process occurs through institutions and protocols that evaluate, disseminate, and reproduce architectural objects. These objects, in turn, shape subjects (see Althusser 1971), positioning them in predetermined social roles. This interpellation happens not only in the theoretical discourse of urban studies but also in the concrete realities of everyday life and the imaginary world of desire promoted by propaganda and media. Thus, denying the architecture of the city's role in reproducing social relations would be, at best, imprecise. Moving from the symbolic to the concrete, from the abstract management of production to the pragmatic experience of everyday life, architecture functions as a means of realizing ideologies in social space (see Spencer 2012, pp. 105–109).

This book asks a central question: why do design practices that aim to deliver radical social transformation fail? The answer lies in architecture's dual reproduction: not only does it shape user subjectivities through its products, but it also shapes the subjectivities of designers themselves—thus reinforcing its own ideological structure, entangled by its history.

This concern is particularly relevant for those searching for a design practice that challenges the status quo. To achieve radical social transformation, one must interrogate not only its products but also the processes through which its subjects are reproduced. This investigation does not propose a new radical method or model; rather, it seeks to enhance critical awareness of stakeholders' positions within spatial production. To this end, it offers a cognitive map, inspired in Jameson (1990)—it's a tool that supports radical practices by identifying key ideological constructs.

Such a tool is helpful because the current debate is mired in false dilemmas and mystifications. The field has become a battleground of stylistic disputes and endless exegeses of canonical narratives. This book, therefore, traces these false dilemmas not to engage in them, but to step outside their cacophony and refocus on a reflexive critique. Reflexivity, in this context, refers to a theory conscious of its own theoretical agency—one that acknowledges its limits while revealing its own self-references.

Thus, this research does not attempt to define the architecture of the city in the abstract. Instead, it critically examines the abstractions embedded within the production of urban spaces and the ideological forces shaping its supposedly neutral concepts. Rather than treating the city as an idealized, pure object, this book investigates how it concretely reproduces abstractions in reality.

The book analyses how urban spatial structures in contemporary neoliberal London produce and reproduce subjectivities. It treats urban space as a dynamic field of intertwined conflicts, employing a three-step analysis: first, examining how neoliberal spatial processes enclose subjectivity through reification (Chaps. 2 and 3); second, analysing how these processes are mediated by design and science through fetish (Chaps. 4 and 5); and third, exploring detours and insurgencies (Chaps. 6 and 7).

By analysing the privatization of public spaces, this study reveals how these changes transform the classical subjectivity of the modern metropolis. If Simmel's classical essay described the modern city as an experience of difference, freedom, and rationality, contemporary neoliberal urbanism instead constructs a post-metropolis defined by domestication, anaesthetic comfort, and invisible cages of control. These spaces do not merely deceive people; they actively frame people's subjectivities around consumerism and individualism.

Furthermore, this book examines policies, planning documents, and scientific discourse to trace how social relations become objectified in urban space through fetish mechanisms. By understanding the political economy behind urban production, we can begin to challenge neoliberal design-thinking strategies.

Finally, this study explores resistance, insurgent practices, and alternative design strategies, including micro-utopias and hacking practices. It aims to open new fields for critical design, urbanism, architecture, and activism—a resource for those seeking alternative ways of imagining the city and developing non-hegemonic design practices.

This is particularly relevant in a time when neoliberal spaces disconnect people from non-hegemonic or collective actions, while subtly controlling urban experience and framing behaviours. This study explores how people come to accept and even love these spaces without recognizing the ideological spectacle at play—one that renders them passive spectators in a landscape of mass dispossession, expropriation, and exclusion.

The topic of 'public space' is complex, and the literature presents divergent approaches. Chapter 2 examines the reification produced by London's privatized public spaces, offering a literature review and a new analytical framework consisting of 36 criteria organized into 5 key aspects. This framework supports a broader theory of urban reification. Additionally, participant observation and interviews conducted in live projects, summer schools, and public space interventions helped contextualize counter-hegemonic practices in response to these changes.

This research follows a Grounded Theory approach, basing the theoretical reflection prioritizing on empirical observation. Instead of seeking to confirm pre-existing concepts, theory is constructed from the recurrent and emergent themes that emerge from the interpretation of the collected data. Consequently, empirical case-study chapters presenting coded concepts precede theoretical discussions (e.g., Chap. 2 before Chap. 3, Chap. 4 before Chap. 5, and Chap. 6 before Chap. 7). Nevertheless, of course, every research starts with axioms guiding interpretation. This book examines how urban spaces reproduce social relations, particularly in shaping subjectivities; and three initial definitions are crucial, as Rossi's radical lesson on the architecture of the city as artifacts is often misunderstood: urban artifacts are produced.

These artefacts thus form historical, cultural, economic, and political complexes materially persisting with a *Tendenza*—a recurring social logic and typological impulse that persists across eras and contexts—directing and shaping the possibilities of life for inhabitants through deeply embedded spatial frameworks (Rossi 1982). Urban artefacts thus actively shape the continuous production of subjectivities. The idea of subjectivities moves beyond individualist perspectives, highlighting polyphonic, collective ways of thinking emerging from interactions among spatial practices, institutions, affects, and social rituals (Guattari 1995). Such subjectivities are never static but unfold within ideological structures embedded in everyday life. These ideological relations operate via interpellation, a process where concrete elements—such as urban spaces, institutions, advertisements, and daily objects—silently impose identities, roles, worldviews, and practices, constituting subjects themselves (Althusser 1971). The architecture of the city thus becomes instrumental in shaping, constraining, and reproducing the subject's position in this world, informing their awareness of the possibilities for agency (Jameson 1990).

Rather than solving or mitigating this issue, this book seeks to reveal how the problem intensifies over time—offering a framework for critical and radical engagement with the production of the city: first and foremost, a subject must be able to position oneself in the world, one must first recognize their own estrangement from the very means of his profession. Thus, we do not seek to "overcome" the fantasies of the discipline, but rather to traverse them—enabling a self-reflexive critique. If one were to argue that this lacks an immediate solution, the counterargument is clear: with the steps taken more significant than finding universal causes are the landscapes they reveal, the paths they open, and the possibilities they enable.

Goiânia, Brazil Camilo Vladimir De Lima Amaral

References

Althusser L (1971) Lenin and Philosophy and Other Essays. Monthly Review Press, NY

Guattari F (1995) Chaosmosis - an ethico-aesthetic paradigm. Indiana University Press, Indianapolis

Jameson F (1990) Cognitive Mapping. In: Nelson C, Grossberg L (eds.) Marxism and the Interpretation of Culture. University of Illinois Press, Chicago, pp.347–360

Rebel Architecture (2014) Al Jazeera Documentary Series

Rossi A (1982) The Architecture of the City. The MIT Press, Cambridge MA and London

Spencer D (2012) Smooth Operators: Architectural Deleuzism in Societies of Control. Unpublished PhD Thesis, University of Westminster

Acknowledgments

I should acknowledge that this book is influenced by a personal journey of displacement, adaptation, and critical inquiry. After first visiting the city and being enchanted by its vibrant and cosmopolitan spirit, the initial inspiration for the research arose from my own feelings of disconnection when I moved to London. As I began living within its urban fabric, I experienced a city whose spatial detachment imposed a subtle yet profound sense upon me—which I initially struggled to identify—somewhere between estrangement and alienation. At the same time, my role as a university professor compels me to confront the challenge of equipping students to engage critically and meaningfully with the built environment. Underlying this academic and professional endeavour were my own personal struggles, and I sought to reconcile a sense of meaningful existence in a seemingly meaningless world. This book is, in part, an attempt to make sense of these pathways.

I am deeply grateful for a fellowship from Coordenadoria de Aperfeiçoamento de Pessoal de Nível Superior (CAPES) and a leave of absence from Universidade Federal de Goiás (UFG), which allowed me to develop this work. I extend my sincere thanks to my supervisory team—Douglas Spencer, Roland Karthaus, Alan Chandler, and Hassan Abdala. Their guidance, critical insights, and enthusiasm were invaluable. I am also grateful to Angioletta Voghera, who, as my director of studies during a NextGenerationEU fellowship, encouraged me to develop this manuscript as a concise framework integrating past and recent research.

I am also thankful to Jane Cheadle, Liesbeth Huybrechts, Peter Princen, Frank Vanden Ecker, Sarah Martens, Jeremy Gilbert, Sarah Wigglesworth, Thanos Zarthaloudis, Lewis Jones, Mark Lemanski, Christoph Hadris, Christian Victor Palmer, Harald Trapp, Mihaela Anca Ciupala, David Mullins, Teresa Stoppani, Giorgio Ponzo, and Georgios Themistokleou for their discussions and suggestions. I also extend my gratitude to Roberto Monte-Mór, Adriana Vaz, Artur D'Aló, Juliana Torres de Miranda, José Cabral Filho, and Marcela França, who took the time to discuss this research in its early stages and during its most challenging moments. And finally, a special acknowledgment to Michela Pace, Anna Minton, and Amir Djalali, whose debates and input significantly contributed to shaping this investigation and deepening my conclusions.

Chapter 2 develops work previously published in Amaral CVL (2016) Private control and public openness: the development of London's public spaces since the mayor's 2009 manifesto. The Journal of Public Space 1:129–146. DOI: 10.5204/jps.v1i1.12, and it also supports discussions on sections 3.2 and 6.1. Chapter 5 extends the approach published in Amaral CVL (2020) The production of project. Ardeth – A Magazine on the Power of the Project 5:56–77. DOI: 10.17454/ARDETH05.05, and it also supports discussions in chapter 7 and section 4.3. Chapter 6 extends the discussion published in Amaral CVL (2024) Studio life. In: Barosio M, Gomes S, Vigliocco E (eds) School of Architecture(s). Springer Nature. ISBN: 978-3-031-71958-5. DOI: 10.1007/978-3-031-71959-2_9; and it also supports discussions in the introduction and section 7.1.

Competing Interests The author has no competing interests to declare that are relevant to the content of this manuscript.

Contents

Chapter 1
The Reproduction of Social Relations by Urban Spaces

Abstract This chapter aims to establish the background discussion about whether architecture can, or cannot, shape social relations. On the one hand, it is common sense that different societies have shaped their cities based on their culture. On the other hand, mainstream architectural theory since the end of the golden age of modernism has dropped any attempt to change social relations through its practices. This chapter defines this paradox in terms of architectural disciplining and its subtle power to unconsciously shape subjects' identities and behaviours, especially when they engage with space distractedly. Thus, the architecture of cities is both conditioned by social relations in the field and sets a dialectics between territory and subjects that frame an image about the world, place subjects in a position in this world, and bound how they can act in the world.

Keywords Urban design · Architecture · Social relations · Disciplining · Dialectics · Interpellation · Autonomy fallacy · Subjects · Reproduction · Critical theory

1.1 Deconstructing Discipline

Approaching Discipline

Architecture and Urban Design as disciplining forces are a social condition that has been produced historically, rather an essential truth. Here, Architecture and Urban Design are approached not as substances (immutable essences or ideals) which could be traced down and defined in absolute terms; rather, they are understood as disciplines: social practices framing ways of behaving, seeing, and acting in the world. Therefore, these terms refer to no distinctive boundaries and the phenomena they encompass are dynamic, scattered, and interpenetrates different realms of social life. Similarly to Henri Lefebvre's account of the term space, what brings its sense together is its accepted notion and the way it mobilizes social practice.

C. V. De Lima Amaral, *The Urban Reproduction of Subjectivities*,
SpringerBriefs in Geography, https://doi.org/10.1007/978-3-032-01465-8_1

> In other words, we are concerned with logico-epistemological space, the space of social practice, the space occupied by sensory phenomena, including products of the imagination such as projects and projections, symbols and Utopias (Lefebvre 1991b, pp. 11–12).

In this context, classical definitions are meaningless for investigating the reproduction of urban space. For instance, there is no point in separating the realms of 'construction' and 'art' in architecture, whereby engineers and common people would produce bad examples of the former, and geniuses and architects would produce the latter. Regardless of who builds or designs, space production always involves the incorporation of habits and, therefore, contributes to the disciplinary reproduction of space of architecture. In these terms, architecture should be viewed through its social dialectics, how it disciplines the practices of spatial production and how its products, in turn, disciplines habits.

The Oxford English Dictionary defines discipline as (1) the practice of training people to—or in the case of a verb, to train someone to—obey rules or a code of behaviour, using punishment to correct disobedience; and (2) a branch of knowledge, typically one studied in higher education. Thus, investigation of architecture as a discipline relates to the two common understandings of this word, emphasizing the social intertwining of both—i.e. how the knowledge of architecture is also the means through which subjects are trained to obey. Therefore, this book is not concerned with an abstract, pure and universal ideal of what architecture 'is'. Rather it concerns an investigation of the concrete experience of contemporary architecture and how it reproduces practices and subjects.

For these reasons, the research departed from concrete fieldwork as a starting point to trace the logico-epistemological elements set in motion in these concrete social practices and how they reproduce and reinforce social relations. Paradoxically, in these terms, what is socially conceived as 'architecture' matters precisely because it is what we aim to deconstruct. This socially conceived idea of architecture and urban design has a social existence and therefore reproduces social habits. In addition, this social existence of urban reproduction has three dimensions—products, methods, and discourses—that correspond to the three means of its reproduction—reification, fetish, and fantasies. Thus, the expression 'urban reproduction' has a triple meaning: urban spaces reproduce social relations; urban spaces are reproduced by specific social actions; and discipline reproduces subjectivities (and the subjects that act on the production of these spaces).

Finally, there is no privileged space for the discussion of high-fashion architecture. This last point relates to the difference between Tafuri's and Kracauer's critiques of architecture. When attempting to radically critique architectural ideology, Tafuri and Dal Co (1976) ultimately reinforced the main ideology of the field—that architecture is the result of individual geniuses. His critique of the history of 'world architecture' is a critique of the discourses and styles of European and American high-profile architects. Alternatively, Kracauer's (1995) essays from the 1920s aim to reveal how the aesthetic act is immersed in a blind field; therefore, what it obscures is more important than what it reveals. In this sense, 'vulgar' architecture—such as the hotel lobby for Kracauer—can more easily reveal the process of reproduction of habits and the blind fields it creates. Later, these 'blind fields' and this reproduction of habits can further elucidate the critique of high-profile architecture (which is here understood simply as the 'core' of architectural ideology).

For a Dialectical Genealogy of the Discipline

The dialectic is contradiction or nothing at all. In the Communist Manifesto, history is contradiction (Marx and Engels n.d. [1848]). Furthermore, the whole effort of Marx (1990) in the first chapter of Capital is to show the fundamental contradictions of the capitalist system: how the reification of social relations contradicts the assumption of 'things as they are' as natural, how 'exchange' value contradicts 'use' value, how capital and concrete exploitation contradicts the supposedly 'free market', and how the labour value embodied in products contradicts the commodity fetishism.

This vital key helps us understand why the deadlocked discourses of architecture are not enough to expose the fundamental contradiction between discipline and dialectics, i.e. the fundamental contradiction between the social apparatuses that discipline architects to behave in a certain way and the dialectical movement to create other possibilities, produced by conflicts, imagination, and desire.

For instance, to say 'architecture is just architecture' is already a contradiction, which states that architecture should be understood as what someone understands as 'architecture'.[1] Similarly, for Marcuse, there is always a struggle, a contradiction in any such affirmation. For example, Marcuse (1967, pp. 133–134) asserts that saying that 'the State is reason' or 'men are free' is already to disregard (or sublimate) all circumstances in which the state is irrational, and men are enslaved. This is, according to Marcuse, an ethical choice, an engaged option that relates particular content to a particular form. Thus, any affirmation already originates from a political positioning in the debate.

Nonetheless, the possibility of contradiction has been denied by a sort of critique founded in absolute negation. For Adorno (1951, §18), 'There is no right life in the wrong one'. For Foucault in Discipline and Punish (1979, p. 84), refusal is only an internal mechanism of the system. Similarly, Baudrillard (1991, pp. 9–14) claims that we are condemned to live in a simulacrum.

Contrastingly, for Jameson (1981, pp. 90–92), these positions within a 'total system' can trace their roots to Weber's 'iron cage', where change is relegated to the

[1] For Lefebvre, to obtain a dialectical logic (instead of an abstraction of the dialectic), one must overcome the old principle of identity (the principle of 'no contradiction' or the 'excluded third', see Lefebvre 1975, p. 2). For Lefebvre, the two terms in a proposition are dialectically implied in one another, transforming each other. This is true even in the case of A = A, where the repetition already introduces a difference (1975, pp. 8–10). Furthermore, Lefebvre argues that connecting form (A) with a concrete reality is already a social and ethical judgement of a fundamental 'fissure' between forms and contents.

'nonmeaningful', thus eliminating the possibility of dissent and reintegrating any form of struggle as a functional part of the system. To avoid 'operational myths' and teleology, Jameson (1981, pp. 139–160) argues for a move towards a genealogy in which one departs from a 'full-blown system' only as a means to retrace history and discover the elements that allowed such a system to emerge.

Furthermore, according to Rancière (1994, p. 59), we must move beyond simplistic 'genealogies': 'a new way of thinking about the transmission of meaning and about legitimate descent'. For him, genealogy is at risk of falling into a field that chronicles the 'nobles'. In this logic of filiations, there is an offspring hierarchy—the firstborn—reducing the diversity of origins into a linear causality that directs to one single cause—the patriarch. In that inverted tree methodology, the present is always subordinated to the past. To overcome this paradox, we should move further towards the original sense of the active nihilism of Nietzsche (1987) and to break with the tree model in genealogy, as Deleuze and Guatarri suggest with the rhizome (Gilbert 2016), which adds the perspectives of horizontal and complex relations of dependency beyond that of offspring. For instance, in the case of the history of machinery, Guattari (1995, p. 40) asserts the following:

> The phylogenetic evolution of machinism is expressed, at a primary level, by the fact that machines appear across 'generations,' one suppressing the other as it becomes obsolete. The filiation of previous generations is prolonged into the future by lines of virtuality and their arborent implications. But this is not a question of an univocal historical causality. Evolutionary lines appear in rhizomes; datings are not synchronic but heterochronic. Example: the industrial 'take off' of steam engines happened centuries after the Chinese Empire had used them as children's toys. In fact, these evolutionary rhizomes move in blocks across technical civilisations.

With those points in mind, what was thought to be irreconcilable, the paradox of a dialectical approach to discipline, can be overcome if one is prepared to accept that no original pure essence of architecture (a glorious lineage) exists and that the social system is not 'closed'; instead, it is a process of 'enclosing' practices by ideology. In this continuous process of enclosure by both the discipline and its products, there is no reason to assume that struggles and 'difference' cannot emerge. Dialectic implies the emergence of difference; and contradicting the idea of an absolute apparatus, the system does not encompass the possible and the impossible.

Can an analysis of architectural and urban design practices in these terms reveal new possibilities for the profession? Can theoretical straitjackets be overcome and a theoretical map for the exploration of new territories of possibilities and struggle be developed?

Towards a Framework for Reading the Urban Social Reproduction

Jameson (2000, pp. 442–461) uses Gramsci to overcome the abovementioned theoretical straitjacket of absolute 'disciplinary apparatuses', by building a perspective that empowers counter-hegemonic strategies to be developed. At a conference in 1990, Jameson (1990) proposed the development of 'cognitive maps' that could help us understand and visualize the current social condition in which we live and thereby place ourselves in a counter-active position in relation to the existing structures. This map could enable a reflexive assessment of the fissures and counter-hegemonic possibilities within the disciplined field. In this sense, even though power and discipline are active realities, they are not pure original essences in society (i.e. part of its fundamental 'nature').

Therefore, to say that there is a possibility of an architecture that struggles against the reproduction of preset fantasies and contemporary social relations is not to say that this is the essence of architecture nor that such architecture exists. Instead, this effort aims to search for this possibility exactly where the disciplinary field insists that it is impossible. Thus, this search implies a wish and, beyond that, announces the ethical motivation for the endeavour to come. However, we do not mean to provide a complete inventory of every aspect of the field. Instead, we aim to construct an instrument to guide analysis and critique.

This instrument can be used to locate the deadlocks that we have identified in the discipline, which can be explored as dialectical conflicts. When all that is solid in the discipline vanishes into thin air, the potential for new possibilities can be articulated. In this endeavour, we have already recognized two forces: on the one hand, the discipline reproducing architecture and, on the other hand, the dialectic and its recognition of the possibility of 'the new' and its appeal to find fissures through which concrete transformations can emerge. To approach the 'discipline', we will first conduct an analysis of the fallacies and dead ends of the discipline at play, only to enable a second step in which to develop a proper dialectical framework to approach the reproduction of architecture.

But first, let's establish some key points. This book is a cognitive mapping of that reproduction of architecture. As intended by Jameson (1990), a cognitive map is a device that helps subjects to navigate in the image they create about their relationship with the world. In this sense, the research aims to develop a cognitive map that enables subjects to reposition themselves in relation to the reproduction of architecture—i.e. the cycle of reification, fetish, and fantasy. Although this condition is inescapable, the development of such a map aims to facilitate critical and reflexive awareness. By unveiling the reification of urban spaces, the fetish techniques used for that, and challenging its common narratives (from subjects to subjectivity), the book poses that we can better situate ourselves in the field and create possible detours, thereby potentially avoiding common traps.

Finally, it is important to note that the triple meaning of the expression 'urban reproduction'—urban spaces reproduce social relations; social action reproduce urban spaces; and the discipline reproduces subjectivities (and the subjects that act on the production of these spaces)—was built on the basis of fieldwork, which investigated both the new spaces being produced in London and design practices attempts to deliver concrete social changes. As the research developed, a series of paradoxes and deadlocks emerged from these experiments on 'radical' architecture. Based on critical observation, the research built a critical theory of how architecture prevents radical transformations of social relations and ultimately reproduces the status quo. Such means of reproduction was (1) 'reification'—the social relations engraved in concrete things—(2) 'fetish'—the methods and means of reification—and (3) fantasies—the narratives that mobilize desires, produce subjects, and discipline subjectivities.

1.2 Fallacies and Dead Ends Reproduced by the Discipline

The Fallacy of Autonomy

Although the fallacy of autonomy is also a type of false dilemma, it has unique characteristics. While the previous arguments of false dilemmas would affirm the possession of a key universality as opposed to a false condition (by building a dichotomy with a straw man), the denial of a conjunction between architecture and society instead amplifies the dichotomy to completely separate the realm of architecture from the world as it is. Thus, only what is maintained within the architectonic *illusio* is viewed as a positive effort, while any contact with reality contaminates the purity of the discipline. This fallacy creates an absolute separation from reality, allowing total freedom to operate in a completely imaginary field. Thus, architecture becomes a fairy tale, developing a series of myths, illusions, internal rules, role play, and so on. In other words, the essence of architecture becomes its own fantasies.

In this sense, the dichotomy presents itself as the recognition of an absolute border: on one side of a line, there is the dirty and imperfect condition of the world (society or related conceptions); on the other side, there is architecture (a crystalline object of purity). Even though this autonomy can be easily unmasked by architectural diversity linked to cultural and historical differences (thus revealing how architecture is connected with culture), this fallacy does not simply assert that the realm of illusions is 'better'. Furthermore, it presents itself as an 'absolute' illusion, more real than reality itself. Thus, it enters the realm of the simulacrum, an illusion built upon illusions.

This is the favourite paradoxical deadlock of famous architects because the inherent circularity of this reasoning reaffirms the central role of celebrities in the 'game'. In what Bourdieu (1996a, pp. 166–173) called 'the production of beliefs', there is an internal 'social logic' in which any field of art determines the value of its products. In a position of celebrities, star architects are given a pedigree in an aristocratic court (Bourdieu 1996a, pp. 11–18), thus evoking a distinguished value to their achievements. This value is considered self-evident, as the existing institutions and structures of the field reinforce it. In other words, the position of celebrity (as part of an architectural aristocracy) is taken as evidence of its true value. In this parallel world, celebrated works are regarded as rules that define what is right or wrong, good or bad, smart or stupid, advanced or backward, and of the future or of the past. In this parallel world, no paradox exists because 'I (the celebrity) am the truth'. In addition, others must attentively respect these definitions.

In this context, much criticism aims to reinforce the parallel reality of architecture. Colin Rowe and Fred Koetter (2000 [originally published in 1978]) and Peter Eisenman (2006 [originally published in 1984]) defended architecture as an autonomous practice that should forget society and concern itself with only 'technical' or internal matters. Architecture thus becomes a geometric 'game'. Shapes from different contexts and epochs can thus be compared without conflict. Even the historicity of geometry can be ignored. The construction of this parallel world allows famous architects to feel comfortable designing for dictators, using construction sites that operate based on contemporary slavery, and promoting primitive accumulation, segregation, and expropriation as if those things 'didn't exist'. Or, rhetorically, 'didn't concern good design', after all, Zaha Hadid was only doing 'design' (understood as drawings) in the case of a dictator's award-winning museum (see Catling 2014; Fairs 2014).

Rem Koolhaas enjoys a specific deviance in this fallacy, resulting in his acknowledged 'cynicism' of the game. His office does not ignore reality completely; it uses reality as a game of representations. For instance, his office plays with history as an imaginary identity, social context as fictional allegories, and laws as rules of the game. The depoliticization of Unger's dialectic (Aureli 2008) allows the Koolhaas 'brand' to instrumentalize the context and irreverently appropriate the most despicable characteristics of the status quo as an inevitable absolute. He can thus develop his work as blind utopias, a kind of dystopia without critique, as he does in 'Voluntary Prisoners of Architecture' and 'Junk Space' or in the transformation of architecture into 'branding' for Prada (see Spencer 2012, pp. 239 and 298).

We will further develop the analysis of this phantasmatic dimension of architecture in Chap. 3. For now, retaining its logical contradiction will suffice: the production of this parallel world of architecture does not avoid the contradictions of society. In fact, because architecture is absorbing and accepting the status quo as inevitable, it fully engages in the process of reproducing current status quo. By proposing the fallacy of autonomy, architecture not only masks its relationship with society but also becomes an instrument that reproduces the fantasies of the status quo (i.e. to accept is already an ethical choice). This process of collective fantasy is only possible through the grounding of the architectural discipline in fetish (the production of facts by means of sprezzatura and techniques of phenomena, as we will address in Chap. 4).

As opposed to the fallacy of autonomy, which implies an absolute border, asserting architecture as an 'automatic' reflection of deeper social causes, i.e. as a superficial image or representation, implies a negation of the dialectic, as we shall see next.

Dead Ends

Dead ends can generally be identified as a lack of critical detachment or a lack of imagination. Nevertheless, the complexity of the problem might blur some boundaries, and it might even be used as a cynical or dissuasive instrument to simulate a radical approach. Nonetheless, these approaches either reinforce existing structures or fail to create substantial change. Dead ends in urban design theory often appear in the following modes: charity, resignation (abdication and abnegation), escapism, and protest.

Charity approaches attempt to solve systemic issues using the same mechanisms that created them. For example, addressing housing shortages through capitalist expansion merely extends the market's reach without altering the reproduction of inequalities. Alejandro Aravena exemplifies this approach, viewing low-income housing as an untapped market rather than a site for structural change. His projects, backed by government and business interests, focus on alleviating disparities rather than eliminating them. In a lecture at UEL (24 November 2015), he explains his proposals as to 'do better' architecture in the very 'same game of the market', and he compared architecture to Brazilian beaches, as a social device that smooths the enormous social contrasts by offering enjoyment and pleasure. When asked if he could draw the line between 'charity' and 'really tackling' the problem, he evaded political responsibility by asserting that he 'only does design'. Ultimately, charity in architecture functions as a façade, mitigating symptoms rather than addressing systemic inequalities.

Resignation manifests in two ways: abdication and abnegation. Abdication denies architecture's social power, treating it as a neutral discipline shaped by external forces. Patrik Schumacher (2015) epitomizes this stance, arguing that architecture should merely reflect prevailing political and economic trends rather than engage with them. He promotes parametricism as the definitive style for the twenty-first century, embracing neoliberal market forces as natural and inevitable. In this view, any attempt to challenge capitalism is dismissed as authoritarian. Douglas Spencer (2012) critiques this perspective, arguing that architects like Schumacher reproduce the mechanisms of a "society of control" (Deleuze). By accepting the status quo as immutable, neoliberalism becomes law, and alternatives are framed as unrealistic. The abdicated architect appears powerless but, in reality, chooses complicity in exchange for professional security.

Abnegation, on the other hand, is a gesture against the self, is the act of denying oneself something. Abnegation does not deny the power of architecture; it instead transfers architectural authority to others, often through participatory design. While Claire Bishop (2006, 2011) acknowledges participation's potential to reimagine social relations, she critiques its depoliticized forms, which have been co-opted into neoliberal and spectacle-driven frameworks. She warns against treating participation as a ready-made solution, noting that it often reinforces dominant structures rather than challenging them. Paulo Freire (1996) similarly argues that the oppressed internalize the logic of their oppressors, meaning that to be genuinely emancipatory, participatory approaches must foster critical consciousness (an awareness of the fabricated relationship between the self and world).

Escapism seeks refuge from societal contradictions through spatial detachment. In the modernist tradition, figures like Gauguin attempted to escape civilization's conflicts by retreating to "primitive" settings. Architectural escapism follows a similar pattern, creating insulated utopias that disengage from structural realities. Oswald Mathias Ungers, of whom Rem Koolhaas is an ambivalent disciple, conceived projects as utopian islands in the city isolated from the regular social structure. Similarly, Temporary Autonomous Zones (TAZs) advocate for transient spaces where social norms are suspended (Galloway 2004, p. 35). These ideas, inspired by Foucault's (1984) concept of heterotopias, prioritize temporary liberation over sustained transformation.

Other theorists propose "micro-utopias"—small-scale interventions that act "as if one is already free" (Sande 2013); Webb and Lynch (2010) analyse how the post-punk scene constructed 'pirate spaces'; Bourriaud (2002a, b) sees these as new forms of micro-politics, as the strategy of total negation and transformation of society is no longer available; John Wood (2007) advocates for imaginative alternatives detached from radical critique and in an ethics of flow. Pier Vittorio Aureli (2013) takes a different approach, proposing a monastic architecture that fosters detachment from society in pursuit of an essential self. Douglas Spencer (2016) critiques his strategies for failing to confront contemporary complexities. By asserting an architectural autonomy detached from historical and social contingencies, they inadvertently align with neoliberal ideology, which promotes the illusion of individual freedom while ignoring structural constraints. These approaches avoid engagement by creating isolated enclaves rather than confronting architecture's role in societal reproduction.

The final dead end occurs when the production of space is reduced to protest. Radical architects often recognize how architecture serves power and ideology but often respond through symbolic resistance rather than systemic change. According to Aureli (2013, p. 56), Aldo Rossi, influenced by Gramsci, viewed architecture as shaped by political economy and social history and sought to expose these dynamics. However, Rossi's transition from critique to proposal led to a deadlock. His designs, such as the Centro Direzionale de Torino, emphasized collective identity but did not alter the mechanisms of architectural production. While monumental in scale, they remained representations of the massive collectiveness of workers, a symbolism expressing a political choice. But it does not interfere with how architecture was produced nor with social processes it was reproducing, and inadvertently it reinforces pre-existing social structures. Architecture, in this model, becomes a static signifier of struggle rather than a tool for transformation. It critiques existing structures without disrupting their reproduction. It cries out for external change rather than actively reshaping conditions. Despite its critical potential, this approach fails to move beyond symbolic gestures.

These four dead ends exemplify architecture's recurring failures to challenge systemic issues. Charity maintains existing structures under a moral guise, resignation surrenders to external forces, escapism retreats from engagement, and protest remains trapped in representation. Each mode ultimately deflects rather than confronts the question of how architecture reproduces society. Moving beyond these impasses requires a critical approach that acknowledges architecture's embeddedness in power relations while actively seeking to transform them.

The Denial of Dialectics

Analysing the denial of the dialectical power of architecture, beyond the simple abdication in the dead ends analysed above, will require entry into much deeper and marshier territory. This trek will demand a preliminary review, a generalization of positions, and a position to be taken in the debate. We will proceed by focusing on the field of critical cultural studies, using the analysis of this problem by Raymond Williams (1973) as a guide and adding contemporary contributions with a special focus on the production of urban spaces.

First, although the increasing complexity and fragmentation of spatial production has transformed 'designers' into a small fraction of the immense machinery of spatial production, 'architecture' has become, as a matter of social fact, an increasingly important device in interfering in the everyday experience of the city, especially in the production of spectacles and simulations of socialization, such as the 'revitalization' of communities. The production of these simulacra reproduces specific social relations in the city (to be further explored in Chap. 2) and have concrete social impacts (to be further explored in Chap. 4).

Nevertheless, some authors present the relationship between society and space as a parallel to the relationship between the base (material conditions, economy, and the mode of production) and superstructure (culture, values, ideas), where the first is the ultimate root of causes and the second is only an effect or consequence of the first. This approach is usually referred to as 'Orthodox Marxism', even though it does not resemble Marx's dialectic approach; therefore, it is often referred to as 'economic determinism' or 'vulgar materialism' (as opposed to 'dialectical materialism'). Nonetheless, many authors have identified this problem and demonstrated the need for a dialectical take on the relationship between the base and the superstructure (Williams 1973; Lefebvre 1975; Harvey 1990; Jameson 1991).

In *The German Ideology*, Marx and Engels (n.d. [1846]) were primarily fighting against ideas and theories that were detached from the concrete experience of reality. For instance, the definition of justice or human rights (e.g. the right to healthy alimentation) had no relevance if it was debated in purely abstract terms. These concepts would only be important according to their actual realization (e.g. what really matters is if people can eat every day). In addition to this concrete use of ideas, Engels (n.d. [1880]), in a text discussing the dialectic method, emphasizes how 'cause' and 'effect' and the limits of things are interactive factors that mutually define one another.

In this sense, the dialectical method does not allow simplifications that establish the 'base' as the determinant of cultural facts. Nevertheless, the mistake might have originated in Engels' [1883] speech at Marx's funeral, when he entered into a brief discourse in emotional and symbolic praise about how his friend brought the economic mode of production to the forefront of cultural discussions. In light of previous observations, this praise does not imply a mechanic determinism of the economic but states the aim of philosophy, which must be focused on the dialectical realization of ideas rather than on an autonomous parallel reality (as we observed previously in the fallacy of autonomy).

Furthermore, the determinism of the base would create a condition in which subjects are only passive spectators of the external world. Such determinism also produces a common deadlock in the architectural paradox, which can be called 'Tailism', i.e. the idea that society will develop itself spontaneously, following the 'tail' of progress. This term was rescued from Lenin by Lukács (2000) in his Defence of History and Class Consciousness. In the context of architecture, the deadlock refers to the belief that the newest style, the newest technology, or the newest social tendency will always be better than the previous one as a result of a 'natural' evolution of humankind. In this sense, the new is accepted as the better or best fit. Here, there is no space for critical practice, nor there is space for human imagination. There is only space for 'sensibility' that can 'capture' a tendency. This reasoning is used by recent 'accelerationists' (Srnicek 2013; Williams 2013a; Mackay and Avanessian 2014) and architects such as Koolhaas (1994 [originally published in 1978], 2002). Lukács was radically critical of this kind of approach because of the mechanical hope in progress and the elimination of conscious subjects concretely participating in the creation of the historical process.

An opposing but mirror-like position is delineated in Manfredo Tafuri's (1976) famous book, *Architecture and Utopia*. Although his analyses are fundamental to the understanding of the articulation between material conditions, the economy and cultural practices, he concludes that architecture is nothing more than an ideological reflection of a mode of production, i.e. nothing more than the appearance of the economic base. Nevertheless, this approach is typical of what has been termed 'Messianic Marxism', where in the absence of the 'Messiahs', a righteous life is impossible.

For Tafuri, architecture is '"Uselessly painful" because it is useless to struggle for escape when completely enclosed and confined without an exit' (Tafuri 1976, p. 181). In this sense, Tafuri conceives of the social 'system' as a perfect enclosed machine. Furthermore, for him, 'the subject is now the system' (Tafuri 1976, p. 55). Based on that conception, in the absence of a communist society, a correct epistemology is impossible. This conclusion cannot withstand Marx and Engels' dialectical approach, and the paradox returns in a new version of the previous deadlock: this time, a form of escapism into the abstract realm of 'negative criticism' and the belief that the critic is a 'free' subject, who is capable of detaching himself from the grime and dirt of reality, thus supposedly bringing a pure 'negativity' to the surface.

In this sense, Jameson (2000) notes four main constraints in Tafuri's thought, which largely relate to his 'negative dialectic': (1) Tafuri conceives capitalism as a completely enclosed system ('far worse than Max Weber's iron cage'); (2) aesthetics is at the same time detached from the 'real' causes of social transformation, and a 'reflection' of structural causes; (3) Tafuri was concerned with revealing 'the truth' of architecture rather than acknowledging that he was producing a truth; and (4) an emancipated architecture can only be conceived in an already emancipated society, only after 'the' revolution has arrived.

Although Tafuri (1976) took important steps in critically demonstrating how different artistic proposals are part of an ideological context—thus implying that any project has a series of social preconceptions, hierarchies, theories, reasoning, causal series, and so on—he ultimately fell into the same abyssal separation of architecture and politics that did Schumacher, as previously cited in this chapter. For Tafuri (1976, pp. 181–182),

> ...it is useless to propose purely architectural alternatives. The search for an alternative within the structures that condition the very character of architectural design is indeed an obvious contradiction of terms. Reflection on architecture (...) cannot but go beyond this and arrive at a specifically political dimension.

This is indeed an obvious 'false dilemma', where things and processes must be 'purely' either 'architecture' or 'politics', which leaves no room for the possibility of a 'political architecture' (i.e. the possibility of politics within architecture and of an architectural practice as political action). In this sense, it is also a deadlock of considering oneself free of ideology and thus suffering from a lack of reflexivity. Only a carefully designed dialectical method can overcome this paradox, acknowledge that any architectural or urban design is political (even if it blindly advocates for the status quo) and thereby be capable of approaching the urban reproduction of social relations.

1.3 Dialectics in the 'Reproduction' of Urban Spaces

In contrast to the abovementioned theories of an 'enclosed system', the reproduction of social relations is dialectical and processual. As Henri Lefebvre argued,

> There can be no reproduction of social relations either by simple inertia or by tacit renewal. Reproduction does not occur without undergoing changes. This excludes both the idea of an automatic reproductive process internal to the constituted mode of production (as system) and that of the immediate efficacity of a 'generative nucleus'. The contradictions themselves re-produce, and not without changes (Lefebvre 1976, pp. 90–91).

In addition, the full disclosure of the concept of reproduction results in a reversal of traditional Marxism. A shift in emphasis from production to reproduction results in a shift in focus from the factory floor to the body, from the realm of work to the realm of everyday life, from the means of production to ideologies; furthermore, it moves the realization of capital from the expropriation of labour to the mobilization of consumption and desire.

In traditional Marxism based solely in Volume 1 of Capital (as argues Harvey 2010), the means of production, the conflict over salaries, the dismal conditions of labour, and the control of the means of production were at the centre of most struggles. In that approach, if the realm of production was the means through which capitalist social relations were imposed, the realm of reproduction was a consequence. Therefore, the realm of the 'reproduction of the worker' was conceived only as the small scale of individual survival related to 'making a living' and 'getting by'—a battle already lost. Thus, the body, the self, the ideologies, the everyday, and the necessities (needs and desires) were understood only as reflections of the production realm. For Harvey (2010), this undermines the whole picture of Marx's 'project', which would concern the 'realisation of capital' in the realm of reproduction in Volume 2.

As is the case with the differences between reproduction and production, the differences between repetition and reproduction are also important. The contemporary intensification of reproduction contrasts with the two common uses of the term: making again—i.e. production in large quantities over time—and imitating—i.e. representation and ultimately simulation. Although these aspects are of fundamental importance in modern times, the intensification of reproduction takes the matter to a new level. While 'simulation' has the same appearance with a false essence, in the contemporary 'intensified reproduction' the appearances change, and the essential social relations are reproduced. While 'technical reproduction' delivers mass production—i.e. the multiple representation of the original, as in a 'photocopy'—in the contemporary 'intensified reproduction', originality functions as a means to hide the artifices of production, where a flexible mode of production makes sameness resemble diversity.

Although capitalist social relations increasingly present themselves as a continuous novelty, under their spectacular appearances lie the old hierarchies, forms of power, and strategies of alienation. In this context, to reproduce is to repeatedly produce the same traps but in different guises. In addition, this reproduction is accomplished not simply by inserting timely elements into a neutral space; rather, the relations of power are imposed via the reproduction of the whole environment. For Lefebvre,

> Power, the power to maintain the relations of dependence and exploitation, does not keep to a defined 'front' at the strategic level, like a frontier on the map or a line of trenches on the ground. Power is everywhere: it is omnipresent, assigned to Being. It is everywhere in space. It is in everyday discourse and commonplace notions, as well as in police batons and armoured cars. It is in objets d'art as well as in missiles. It is in the diffuse preponderance of the 'visual', as well as in institutions such as school or parliament. It is in things as well as in signs (the signs of objects and object-signs) (Lefebvre 1976, p. 86).

To clarify this argument, Lefebvre uses architecture as an example. He argues that 'When architectural urban space responds to the "social commission" of developers and the authorities, it is contributing actively and openly to the reproduction of the social relations. It is programm[ing] space' (Lefebvre 1976, p. 88). Furthermore, Walter Benjamin (2008, pp. 36, 40), in his essay about the work of art in the age of technical reproduction, argued that architecture not only provides an object that is collectively received (through perception and only visually). Moreover, architecture is absorbed by the masses in a state of distraction because buildings are received through use (tactilely). Through use, architecture creates habits: tacitly framing tasks and engaging actions. In this way, through designed space, specific social relations become incorporated habits.

Challenging the Reproduction of Urban Spaces

Henri Lefebvre is a philosopher of great importance in the theory of space. As such, he has particularly influenced disciplines such as geography, urban theory, and architecture in recent decades. A prolific writer, Lefebvre wrote more than 60 books on various topics, and even though he has been very involved in the architectural scene (Stanek 2011, pp. 33, 40), none of his work was known to specifically address architecture until 2008, when Lukasz Stanek retrieved a lost Lefebvre manuscript entitled *Towards an Architecture of Enjoyment*, which was written in 1973 and published only in 2014 (see Lefebvre 2014). Writing specifically about architecture, Lefebvre outlines his view on the possibility of thinking about architecture in revolutionary terms.

Lefebvre asks, '…an architectural revolution? Why not?' Although Lefebvre (2014, p. 27) argues that architecture has the potential for achieving 'radical subversion', setting 'aside' relations of production and temporarily 'turning the world upside down' (as would occur within large institutions), he also admits that '[t]his project alone is incapable of changing the world'. As he has argued elsewhere, a revolutionary project needs both 'contempt' and 'corruption'. In other words, it needs, on the one hand, a form of consciousness that critiques a social condition and, on the other hand, a form of consciousness that can also somehow result in the concrete 'rehabilitation of everyday life' (Lefebvre 1971a).

Thus, his curtailment of the discipline is a means of narrowing a form of provocation: 'Can change occur without expectation, without exploration of the possible and the impossible?' (Lefebvre 2014, p. 27) Essentially, his argument is that although a revolution produced by architecture alone is impossible, no revolution is possible without a renewed imagination of social spaces. In other words, there is no architectural revolution, and there is no revolution without architecture. For these reasons, his inquiry (in the abovementioned book) aims to go beyond 'the architect', 'the discipline' and its internal disputes to approach architecture in a wider sense: 'It is a question of "mankind" and its future' (Lefebvre 2014, p. 29).

Therefore, Lefebvre's initial question ('[A]n architectural revolution? Why not?') has conflicting connotations. It can be read as a rhetorical device, a challenge, a provocation (negative or positive), or the perplexity of a disbeliever. The irresolute character of this intriguing question might be caused by the different levels upon which it acts—what it enunciates, what it desires, and what it implies.

With that question, Lefebvre is simultaneously enunciating a possibility, desiring a challenge, implying a fact, and considering a condition sine qua non. Later, he describes the revolutionary possibilities for architectural practices but also the limits of architecture. He imagines how a revolutionary architecture can exist by producing virtual objects (of desire) but implying the constraints in his questions: Why not? Why isn't architecture revolutionary? This impediment often arises because the space in which architecture operates is not a neutral space. Space is already a social product.

For Lefebvre (1991b), space is not an a priori element of reality nor an abstract entity into which one can insert contents; instead, space is socially produced. For him,

> To speak of 'producing space' sounds bizarre, so great is the sway still held by the idea that empty space is prior to whatever ends up filling it. Questions immediately arise here: what spaces? and what does it mean to speak of 'producing space'? (Lefebvre 1991b, p. 15).
>
> Every social space is the outcome of a process with many aspects and many contributing currents, signifying and non-signifying, perceived and directly experienced, practical and theoretical (Lefebvre 1991b, p. 110).
>
> [...] if there is a productive process, then we are dealing with history; [...] we may be sure that the forces of production (nature; labour and the organization of labour; technology and knowledge) and, naturally, the relations of production play a part (Lefebvre 1991b, p. 46).

Therefore, as space has a changeable historical and cultural existence, it is not a simple thing; it instead encompasses the (re)production of social relations. For Lefebvre,

> The object of interest must be expected to shift from things in space to the actual production of space (Lefebvre 1991b, p. 37).
>
> (Social) space is not a thing among other things, nor a product among other products: rather, it subsumes things produced, and encompasses their interrelationships in their coexistence and simultaneity – their (relative) order and/or (relative) disorder. It is the outcome of a sequence and set of operations (Lefebvre 1991b, p. 73).
>
> What exactly were the great cathedrals? The answer is that they were political acts (Lefebvre 1991b, p. 74).

Lefebvre's theory defines space as a product of social, philosophical, and practical human activities; therefore, it is not a natural fact but a social construct with social implications. Furthermore, for him, the production of space is a key element in the reproduction of social relations in general. In the book, *The Survival of Capitalism: Reproduction of the Relations of Production*, Lefebvre (1976, pp. 83–86) argued that capitalism has managed to avoid extinction through the production of space. According to his argument, because space is produced within a cultural logic (say, capitalism), it also reproduces a field of everyday life in which rules and social relations are objectified in space. In this way, capitalism reproduces itself as the dominant social reality, and architecture plays an active role in this reproduction.

> An even more significant illustration of this can be obtained from architecture, a specific, partial and specialized practice which has close links with the everyday. The Architect receives what might be termed a social commission, forcing him to realize spaces which suit society, i.e. which 'reflect' its relations by concealing them under the décor (Lefebvre 1976, p. 88).
>
> They mould it, fill it, and produce their own specific kind of space, which is both homogeneous and fragmented, visual and pulverulent. The architect cannot free himself from them, either in practice (his projects and designs) or in his imagination. (...) But it is the everyday that carries the greatest weight (Lefebvre 1976, p. 88).

> Power occupies the space which it generates, the everyday is the very soil on which the great architecture of politics and society rise up (Lefebvre 1976, p. 89).

For instance, 'private spaces' become a social reality when spatial elements (which distinguish them from public spaces) externalize abstract concept onto concrete space. Thus, spatial elements (say, fences, limitations, contracts, walls, borders, coercive control, and so on) are both concrete and abstract.

In this sense, architecture as a social practice is not simply inserted into the general logic of society. In a society ruled by commodity fetishism and alienated social relations, architecture becomes a device that continuously reproduces that logic within space. Thus, architecture reinforces and traps society inside a restricted set of rules, thereby reproducing society. It is in this sense that the expression 'the urban reproduction' have that triple meaning: it reproduces of society; it is reproduced by society; and it reproduces its subjects. Nevertheless, for Lefebvre, this social reproduction is not 'a pre-existing system'. Therefore,

> ...it follows that the 'real' cannot be enclosed. It is not a situation where there is no possible outcome, nor is the only outcome global collapse; for the contradiction themselves develop though unevenly. And finally, theoretical concepts may escape the system, even though they are born in it and have emerged from (...) space dominated by the strategy of homogenization and of the programmed everyday. But they still have to free themselves from that system (Lefebvre 1976, pp. 90–91).

In this sense, our findings should be differentiated from current use of Foucault's idea of the 'apparatus' (*dispositif*). Clare O'Farrell (2005, p. 129) has defined an apparatus as 'various institutional, physical, and administrative mechanisms and knowledge structures, which enhance and maintain the exercise of power within the social body'. This is an important point to consider when analysing how the designers discipline frames the production of spatial forms and how it reifies subjectivities (of the public but also of the designers themselves) to reproduce itself, trapping reasoning within its rules. However, to understand society as an apparatus or a 'trivial machine' misses the point that society is neither an absolute totality nor 'total' stability. Social space is not a 'trivial machine' in which an input inserted into one side results in a precise and controlled output.

Von Foerster (1972), discussing trivial and non-trivial machines, used physics and philosophy to create what came to be known as second-order cybernetics. As we can see in the diagram below (Fig. 1.1), he argues that a trivial machine is characterized by a one-to-one relationship between input and output, where this invariability precisely defines what a trivial machine is. As a deterministic system, it is predictable. However, in the non-trivial machine, the input-output relationship is variable and depends on previous inputs. A broader description of this relationship can be described in the second algorithm below, where the input, once processed, produces a transformation in the structure of the machine and the output creates a new context that further transforms the internal structure of the machine. Thus, the non-trivial machine operates with conflicting internal structures. This new structure produces a virtual transformation of the original context of inputs, further conditioning subsequent inputs. Thus, the concept of a non-trivial machine can

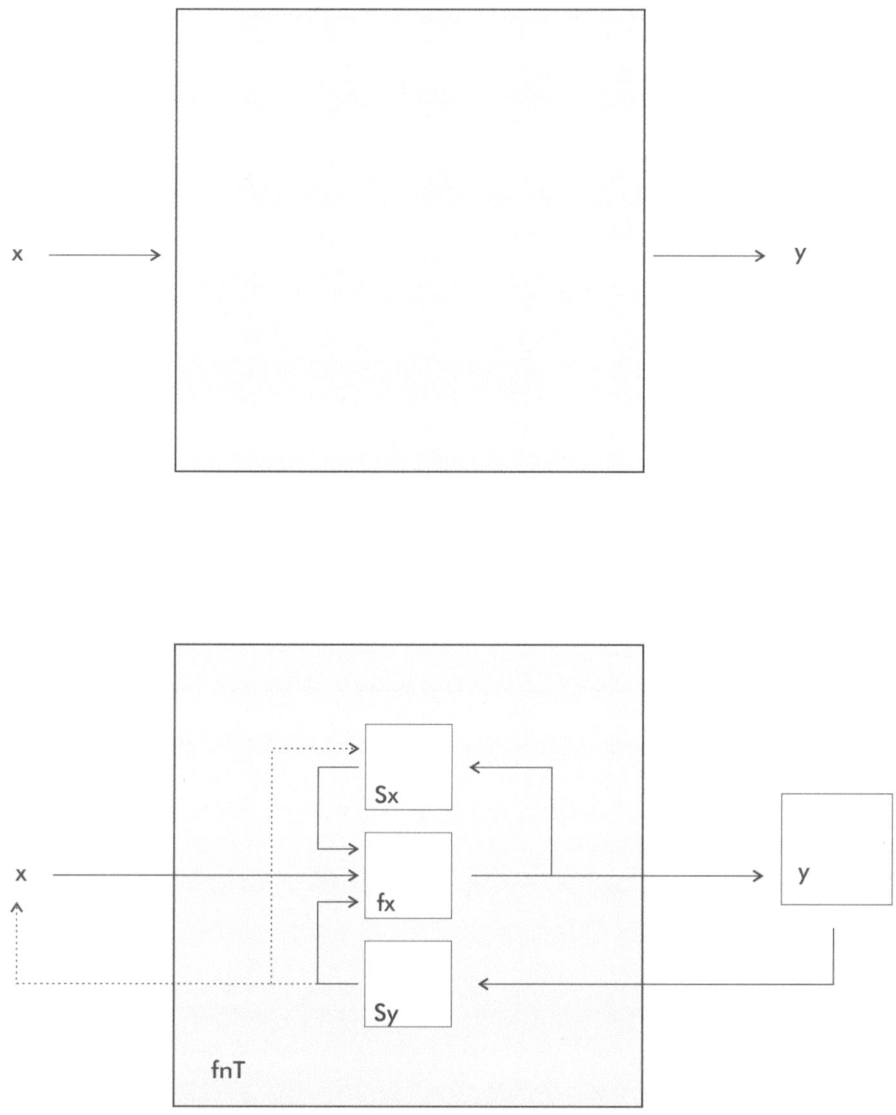

Fig. 1.1 Trivial and non-trivial machines. (Source: the author (based on—but not identical to—ideas from Von Foerster 1972))

elucidate how a social space operates in a complex field of historical and conflicting relations.

Furthermore, to discuss concepts of subjectivity, Alberto Toscano (2006, pp. 136–156) introduces dynamic system theories, arguing that a system can be conceived without a previous hierarchical order, with complex interaction between the internal and the external, and with no stable and distinct objects. He relates this idea with various elements of the theory of the individual in Simondon and discusses the idea of the 'theatre' of operations and individuation. In the process of producing a metastable system, irreversibility is created, which he calls 'trace' and which determines the context of the next transformations. This conception aims to avoid genetic determinism (as in biology) and representationalism (as in psychoanalysis) (Toscano 2006, p. 148). It also avoids deterministic theories, rejecting the possibility of isolating categorical causalities, but without falling into the traps of holism—for instance, as in Spinoza, where everything is God—and organicism—where everything has a pre-established place—in which no transformation of a given system is conceivable (Toscano 2006, p. 149).

Only in this way, we can approach the disciplined system of urban reproduction not as the product of a sort of 'Big Brother conspiracy' and can understand it as a metastable field in which conflicts and tensions can emerge dialectically. In this sense, architecture is not simply the reflection of a status quo although it operates in the reproduction of specific 'traces' and in the permanence of specific social structures. And it is these specific structures that have to be deconstructed. In these terms, the idea of 'reproduction' is a key concept that is useful in overcoming the paradox of 'to shape or not to shape': architecture is reshaping and keeping at the same time, over and over again. Therefore, the concept of reproduction leads to a continuous conflict between disciplining forces and dialectical struggles. Nonetheless, conflicts are what dialectics is all about.

Chapter 2
The Privatization of Spaces in London and the Metropolis Enclosure

Abstract This chapter analyses case studies about the privatization of public spaces in London, by tracing how they change the experience of the metropolis. If the modern metropolis was once read as the space where people would act in a rational, detached, and free manner, now, by analysing these new spaces it is possible to see how they act on affects by framing experiences in terms of individualism and consumerism, and by delivering complex sets of rules that subtly controls users to become passive and obedient. This chapter also traces this process back in time and compares it with the common land enclosure acts to understand how it sets new neoliberal social relations. Finally, by refuting the main argument in favour of the privatization process, it traces down a basis of how urban artefacts are objectified social relations, to be analysed in the next chapters.

Keywords Privatization of public spaces · Metropolitan enclosure · Neoliberal urbanism · Reification of space · Affects · Spatial subjectivity · Driven spatial consumerism · Tangled control · Velvet architecture · Urban artefacts and social relations

2.1 Analysing the Privatization of Public Spaces in London

For Neil Smith and Setha Low (2006) public spaces have a specific political role in modern society. They argue that the definition of Public Space has historically been constructed as the opposite of Private Space. Therefore, both social constructions are the result of modern capitalist society. In this sense, this phenomenon is the result of social struggles against the former feudalist spatial order, and it is even a progressive development from the Greek polis, which did exclude both slaves and women from its political space, the Agora. Thus, public space arises as a new set of social relations, articulating the power of civil society, the market and the modern State (Smith and Low 2006, p. 4). That sets quite a few challenges for a scientific investigation on public spaces.

The relationship between science and reality has evolved through various philosophical and sociological frameworks. Debates over the legitimacy of theories

C. V. De Lima Amaral, *The Urban Reproduction of Subjectivities*,
SpringerBriefs in Geography, https://doi.org/10.1007/978-3-032-01465-8_2

reflect deeper tensions within rational thought itself (Popper 1989; Feyerabend 1989; Adorno and Horkheimer 1996; Santos 2003; Guest et al. 2011). Beyond epistemological concerns, science has also become entangled with power structures, shaping nature and social life through its own sociological dynamics (Kuhn 1987; Bourdieu 1996c; Santos 2007). In this context, the validity of a theory increasingly depends on its effectiveness in mobilizing attention, authority, or dominance within the global division of scientific labour.

Bourdieu (1996a) analysed how fields like art or science function through internal rules, hierarchies, and legitimized 'habitus' that reinforce dominant ideologies. As Santos (2001) reminds us, every interpretation—such as of public space—is socially engaged and departs from the status quo. Likewise, other authors argue that reality is socially constructed, shaped by deep histories and epistemic regimes (Escobar 1996; Lefebvre 1967; Rancière 2005; Santos 2007). This produces a paradox: while tradition weighs heavily on our capacity to interpret public spaces, the only absolute limits to inventing new approaches to it is its social resonance and endurance.

In response to these challenges, this research adopts a methodological pluralism and an epistemic openness, combining Grounded Theory, Thematic Analysis, Participant Observation, Semi-Structured Interviews, and academic debates, theoretical reflection and critical analysis. Following the "constant comparative analysis" method from Grounded Theory (Glaser and Strauss 1967; Guest et al. 2011), data collection and theoretical reflection occurred simultaneously, with emerging codes and themes continually tested against new observations. This set allows empirical evidence to inform conceptual development and offers insight into how privatized public spaces operate as lived phenomena.

Pragmatically, the research followed three main stages. First, a quantitative analysis of planning documents and online information assessed how recent public space designs in London incorporated principles of privatization. Second, a literature-based framework defined the expected features of public space and the mechanisms of private control, guiding the qualitative fieldwork. Third, field observations were coded into patterns of behaviour and recurring characteristics. These codes were grouped into three thematic categories—the dimensions of Velvet Ground, Tangled Orbits, and Spatially Driven Consumerism. Below we present some first impressions on these spaces, so to illustrate how they operate in concrete experience and contextualize the reader. Later we present the first phase observations; the second phase coding that generated the three dimensions mentioned; and a final argument about the critical importance of acknowledging the specific aspects of these privatized public spaces.

The Reified Aesthetics of Privatized Public Spaces

Beyond architecture simply being appropriated by distraction—as in Benjamin's (2008) famous theory mentioned in Chap. 1—it also functions here to create both smooth flows and striated paths. Therefore, we need an aesthetic narrative capable of capture these dialectical dimensions of reified spaces. To exemplify this newly observed spatial experience, we initially propose a strategy of representation that goes beyond the simple collection of data. This strategy seeks to convey the experiential atmosphere in a manner similar to Baudrillard's *America* (1988), recounting the routes and detours of these spatialities. The aim is to rupture—as Baudrillard would put it—the spectral form of ideological appearances, the emptiness of its signs, the seismic shapes, the fractal and interstitial culture. By shifting away from the world as it is, we reach the fragile, tactile, mobile, and superficial culture. This vertigo—always in the edge of collapse—allows to synthetize the whole in each part, revealing the spectral, the fake serenity, and the soft resort-like simulations of a civilization. Neither dream nor reality, the first three examples of these enclosed spaces will serve to illustrate this new landscape, like frames from a film. Thus, Figs. 2.1, 2.2 and 2.3 aims to highlight how the different elements work together in the production of this spatial experience.

Number 1 New Street Square seems like a public space, but it is entirely privatized. Walking down the street, the development boundary is easily perceived. Predominantly grey, the cluster of buildings has gaps in its corners, like spatial

Fig. 2.1 Film frame 1: Number 1 New Street Square. (Source: the author)

Fig. 2.2 Film frame 2: The Aloft. (Source: the author)

Fig. 2.3 Film frame 3: The Renaissance. (Source: the author)

fissures that function as portals—at once a transition and a passage through the complex. A small totem with chrome maps, information, and a CCTV drawing subtly marks the entrance of this urban transition. Everybody is wearing a suit, some passing by in a hurry. Two figures in black stand in front of different buildings. As I walk around, I imagine that one of them is watching me, so I take pictures with caution. On the other side, in a calmer street, the boundaries become complex, with public and private limits meandering through a valley of bollards. That creates a sort of invisible barrier, also enforced by subtle shaded lines in the pavement that swing around like a snake, lassoing detached urban furniture, such as a bench where a single working-class figure is seated, maybe hoping to be in a public part of this estranged space. Entering the square, many people are seated in cafes, but only one person is seated on the square benches. In Land Securities website own words, the place is thus outlined: 'There has been a real sense of magic around New Street Square since it opened, with the dramatically expanded public realm being used for everything from live sports screenings to local food markets'. I sit there for a while, but feel awkward, as if I am not supposed to be there—as if a lively event is not happening at place to host you, you do not feel at the right place. I considered buying a coffee, so I could spend more time in the space, but I decided to face the awkwardness and spare some change.

The Aloft development is also a highly privatized space that seems public. To arrive at the hotel in London's Docklands, pedestrians coming from the DLR must walk through a suspended platform, fluidly connected to the station and the Excel exhibition centre. A long, smooth perspective connects the entrances without any stairs. Along the way, pedestrians can see a square underneath. A profusion of signs is attached to lampposts, with information about its specific rules that no one reads, although it creates a sort of awareness that something different is going on, as the CCTV drawings keep reminding you that you are being watched and you regulate your behaviour as you sense you're being controlled. The myriad of bollards create boundaries without barring you, composing an infinite-like spatial choreography. In the same sense, benches seem to dance out of cracks in the pavement (certainly a security-by-design barrier against terrorist attacks). The place appears overdesigned, as if the invisible hand of the architect is directing each footstep you take. Like jewellery on display, the place seems more an object to be desired and contemplated from a distance through the DLR windows, as its arid and vast dimensions are not very welcoming to be experienced.

The Renaissance development, although not public, does not feel overtly privatized in general appearance—only in its details. Coming from the station, before arriving at the place, we first reach a square, passing through old brick arches. This reminiscence creates a tactile historicity, warming up the senses as we approach the colourful glass and metal building. The formlessness of the square is subtly coordinated by a grid of squares in the pavement, giving command to what otherwise would be an interstice. As in other places, the snaking lines in the pavement are lassoing urban furniture and objects, twisting pedestrians' trajectories from one side to the other. Although the furniture feels more welcoming, it is also an omnipresent gesture of design. While pins in benches avoid skaters, massive totems announce information and the private/public partnership, nailing their sense of place into the ground. As in many other new developments, Renaissance is a mixed-use complex. In front of the cultural building, a poster—from the Lewisham Borough in partnership with the Fusion Management Organization—proudly announces: 'The great outdoors Now even greater', as here citizens can do exercise in the backstreet park, but only by paying for the group classes being offered.

These new spaces illustrate recurring trends in corporate public spaces. They homogenize difference, transform relations into a controlled game—a device for regulating diversity and directing social behaviour towards consumption. Therefore, citizenship is reified into consumerism. Thus, it creates a fundamental change in the experience of the city, directly interfering in how it can be appropriated and in the freedom with which it can be experienced. Without a constant frame of reference, the multiple-layered norms of these spaces create a constant metamorphosis in how citizens are expected to behave, generating anxiety through their unclear urban rules. Thus, these spaces are shaped by the dissolution of the rational, abstract, and conflictual conditions of the old modern metropolis (see below discussion on Simmel's approach to the modern metropolis), provoking the accommodation of differences into a new continuum of heterogeneity in sameness. Furthermore, their complex, knotted and twisted rules differ from former means of control (panoptical and gridded). These places look cosy and kind, domestic and comfortable—as long as you follow their guidance. Rather than disciplining and punishing, this new form of control operates first through disorientation and then refuge in formatted experiences.

First Phase Observations

The first phase of the research aimed to understand the scale of the privatization of public spaces in London. It was accomplished through an initial inventory conducted by combining information from various entities.[1] The focus of case studies was on privatized public spaces approved and developed in London after the 2009 mayor's 'Manifesto for Public Space' (Johnson, 2009) until 2015. From an initial list of 1024 projects, 708 developments were found to be already completed, of which 101 projects fit the criteria of having been approved after the mayor's first policy documents were drafted. Of those, 25 were found to not enter the realm of public and private boundaries (being mainly internal refurbishments). Thus, 76 cases were initially assessed through: the public realm impact; the public space privatization; and the resulting public space's 'coefficient of conversion'. The

[1] Including the planning portal of the UK government, the Greater London Authority website, the British Property Federation website, the New London Architecture organization, and the websites of major property corporations and local authorities. Each one of the 32 boroughs and the Corporation of the City of London have different websites with different search engines, which means that the information is neither transparent nor accountable.

average of these three analyses created an overall coefficient of privatization of the intervention in public space.

The following diagram 32 represents the degree of privatization in these spaces. The first bar represents the scale of the public realm impact: 31 projects (40.8%) impacted on the surroundings (red); 31 projects (40.8%) made small impacts (yellow); and 14 projects (18.4%) had no impact on the openness of the public spaces (green). The second row represents how much the public space was privatized: 42 projects (55.3%) privatized more than 2/3 of the public spaces (red); 27 projects (35.5%) privatized between 1/3 and 2/3 of the space (yellow); and 7 projects (9.2%) did not privatize any aspect of the public spaces (green). The last bar represents the sum of the two indexes (for each project), resulting in what we called the public space's 'coefficient of conversion' into privatized public spaces. The average of these three analyses creates the overall coefficient of the reification of the public space, diagrammatically represented in the map of London (see Fig. 2.4).

Those first results led to the consideration that the reification in contemporary metropolitan spaces demands further dialectical analysis, and that can only be understood by the dialectic between the physical arrangements and the immaterial experience (including economic, political, and social aspects). 'Things' in public space are not a matter of simple barriers, but how the objectified social relations reproduce citizenship roles in the space, and how social codes and interpersonal rules are engraved in spatial elements. This creates a new spatiality, where citizens move across it in unconventional dimensions, thus, we will need new concepts to understand its multiple aspects and tensions (to be done in the next section).

2.2 Second Phase: Excavating the Reified Urban Experience

A second round of qualitative analysis aimed to reveal the mechanisms through which urban spaces are creating specific forms of social relations. These new forms produce a similar effect to the process of primitive accumulation, in which the citizen is progressively separated from the means of free place appropriation at the same pace at which reified behaviours are imposed. This procedure was made possible by strategies that hide the commodification of place and create a sense of a 'naturally artificial' social order.

For Approaching the Publicness of Space

The reproduction of social relations by the privatization of public spaces is a complex aspect to observe, as it is a diffuse form of control. It is also a difficult aspect to assess because there are different means to influence behaviour and articulate relations between people in space. Furthermore, there is a conceptual difficulty in defining what public space is and how it relates to the promotion of social relations.

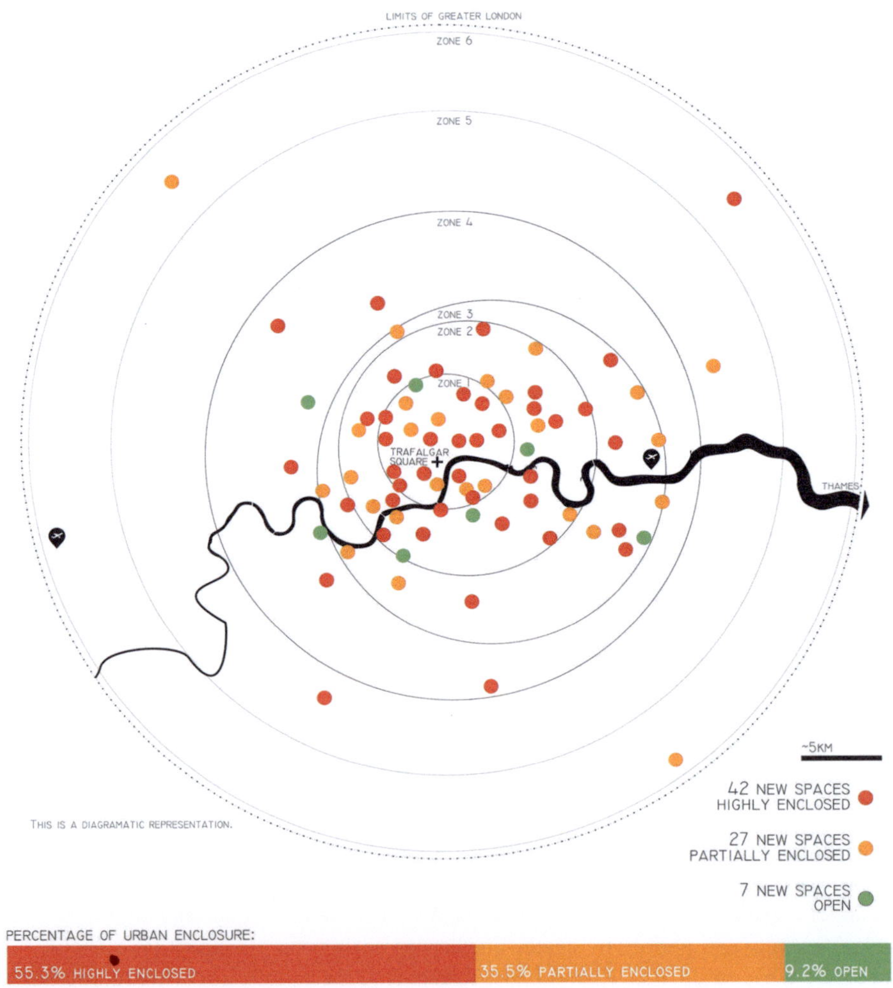

Fig. 2.4 Public space enclosures in London in a non-georeferenced schematic diagram. (Source: the author)

The literature presents multiple methodologies to assess the 'publicness' of a place into divergent grids and concepts (Varna and Tiesdell 2010; Benn and Gauss 1983; Kohn 2004; Németh and Schmidt 2011; Németh 2012; Lessig 2001; Staeheli and Mitchell 2008; Iveson 2007). For this second phase, a form was developed to combine different field techniques and to observe spatial apparatuses. This form was created by a critical synthesis of the many methodologies to measure the 'publicness' of spaces, resulting in 36 criteria to be checked. These criteria and considerations derived from the following studies:

1. Varna and Tiesdell (2010) characterized the conceptualization of publicness in two major effort groups: (a) *Inductive*: the quality of being public comes from an external element, physical properties, material elements and barriers; (b) *Deductive*: comes from internal issues, symbolic interpretations, rules, socially constructed meanings (in the eyes of the beholder). They propose a model for publicness interpretation by measuring ownership, physical configuration, control, civility, and animation.
2. Benn and Gauss (1983) propose three dimensions for the publicness examination: (a) *Access*: the ability people have to occupy the space, i.e. appropriate the space and recreating its function; (b) *Agency*: the examination of who defines how the space is controlled, i.e. if the rules have social accountability; and (c) *Interest*: the examination management decisions regarding the space, interpreting if the 'rules' benefit the public or particular interests.
3. Kohn (2004) uses three criteria for publicness: (a) *ownership*; (b) *accessibility*: the freedom, control, and neutrality of access; and (c) *intersubjectivity*: how the elements of the urban space facilitate or make difficult the encounter and interactions of people, allowing difference to emerge.
4. Németh and Schmidt (2011) has used three axes to analyse publicness: *ownership, management*, and *uses/users.*
5. Staeheli and Mitchell (2008) argued that a space is truly public when it can integrate the inhabitants into the creation and use of the space. For that, it is important to analyse the receptivity, welcome, and comfort provided.
6. For Iveson (2007) spaces becomes public through a process that engages people in debate of the space's proprieties itself. Therefore, a place is public when it is subjected to public assessment, or if it becomes a means of *public conscience.*
7. Lessig (2001) creates a model of interpretation of space creating an interpretation grid drawn from information network society. He uses three 'layers' for that: (a) *Physical*: the hardware is the material elements and infrastructure (cables, mediums, objects); (b) *Code*: the laws, codes, protocols, and rules that mediates the exchanges; (c) *Content*: the actual information and elements involved in the transaction between people.
8. Németh (2012) asserts that any 'truly public forum is characterized by (relatively) open access, unmediated deliberation, and shared participation (...) Nevertheless, publicness is always subjective: whereas some might feel a space full of homeless persons is "truly public", this sight might drive other users away. (...) A simple metric might examine public space vis-à-vis' [a] *Free Access* and [b] *Behaviour Freedom.* He suggests that a complete analysis should con-

sider the following aspects: Physical (spatial programming, mobility, restrictions, and aesthetic); Code (laws, regulations, policing, norms, language, and guidelines); content (use, behaviour, symbolism meaning).

9. For Zukin (2010) culture used to be place-bounded, and Public Space was connected to its history, but it is not anymore. Rather than preserve only the buildings, Public Spaces should also preserve the community that created the authenticity of the place. Domestication by cappuccino is a form of controlling the space by consumption taste. She argues one should not lose the big picture: the issue is about capital granting incomes, controlling workers, associating with state power to make profitable rules, and using media apparatuses for managing consumer tastes. For her, authentic democracy is loud, unruly, unpredictable, dangerous, undisciplined, independent, and non-programmed behaviour. Therefore, Public Spaces should be free for Protest, as the political aspect is one of most important elements of the polis.

This review resulted in a multi-layered methodology registering different aspects of the spaces under consideration (see Fig. 2.5). This aimed, on the one hand, to deliver a rigorous analysis and, on the other hand, to allow flexible observation of the regular experiences of common users. In this sense, the case studies were a mixed methodology involving dialectical investigation, analytical techniques, behaviour and aesthetic observation, and critical synthesis. The observations of each case study were than filled in the form divided into five main aspects: (1) Descriptive and informative elements; (2) Territoriality aspects: observation of how the object shapes identity, flows and boundaries in relation to its surroundings; (3) Mechanisms of Heteronomy: analysis of how signs, furniture, and spatial devices guide behaviour and structure experience; (4) Observation of Public Experience: assessment of how elements of affect, events and actors interpellates subjects and suggests roles and actions; (5) personal experience: synthesis of the site through critical and aesthetic photographic interpretation.

In this second phase of the assessment, 20 spaces (including the three developments mentioned above) were selected with different private/public arrangements. The combined analysis of textual documents, data collection, field qualitative interpretation, and theoretical reflection made it possible to formulate a grounded theory of the contemporary production of reified urban forms. This analysis of the new public space strategies in London allowed to develop an interpretation of the mechanisms through which the urban space is being enclosed to be discussed in the next section.

The Three Dimensions of Enclosed Public Spaces

These new urban forms amplify ideologically driven behaviour, moving it from production to reproduction. Thus, the reification of the city's spaces extends its focus from the workspace (the organization of the factory into an assembly line, for

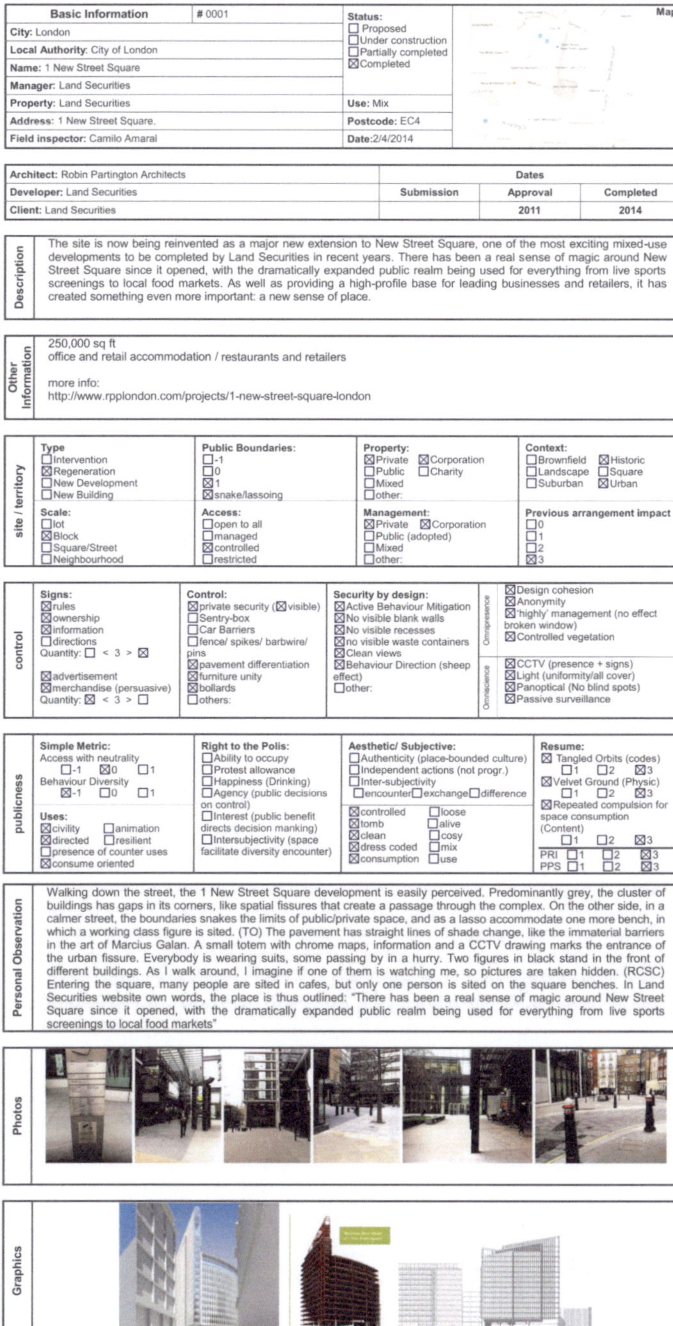

Basic Information	# 0001	Status: ☐ Proposed ☐ Under construction ☐ Partially completed ☒ Completed	Map
City: London			
Local Authority: City of London			
Name: 1 New Street Square			
Manager: Land Securities			
Property: Land Securities		Use: Mix	
Address: 1 New Street Square.		Postcode: EC4	
Field inspector: Camilo Amaral		Date:2/4/2014	

Architect: Robin Partington Architects	Dates		
Developer: Land Securities	Submission	Approval	Completed
Client: Land Securities		2011	2014

Description
The site is now being reinvented as a major new extension to New Street Square, one of the most exciting mixed-use developments to be completed by Land Securities in recent years. There has been a real sense of magic around New Street Square since it opened, with the dramatically expanded public realm being used for everything from live sports screenings to local food markets. As well as providing a high-profile base for leading businesses and retailers, it has created something even more important: a new sense of place.

Other Information
250,000 sq ft
office and retail accommodation / restaurants and retailers

more info:
http://www.rpplondon.com/projects/1-new-street-square-london

site / territory

Type	Public Boundaries:	Property:	Context:
☐ Intervention ☒ Regeneration ☐ New Development ☐ New Building	☐ -1 ☐ 0 ☒ 1 ☒ snake/lassoing	☒ Private ☒ Corporation ☐ Public ☐ Charity ☐ Mixed ☐ other:	☐ Brownfield ☒ Historic ☐ Landscape ☐ Square ☐ Suburban ☒ Urban
Scale: ☐ lot ☒ Block ☐ Square/Street ☐ Neighbourhood	**Access:** ☐ open to all ☐ managed ☒ controlled ☐ restricted	**Management:** ☒ Private ☒ Corporation ☐ Public (adopted) ☐ Mixed ☐ other:	**Previous arrangement impact** ☐ 0 ☐ 1 ☐ 2 ☒ 3

control

Signs:	Control:	Security by design:	Omnipresence / Omniscience
☒ rules ☒ ownership ☒ information ☐ directions Quantity: ☐ < 3 > ☒ ☒ advertisement ☒ merchandise (persuasive) Quantity: ☒ < 3 > ☐	☒ private security (☒ visible) ☐ Sentry-box ☐ Car Barriers ☐ fence/ spikes/ barbwire/ pins ☒ pavement differentiation ☒ furniture unity ☒ bollards ☐ others:	☒ Active Behaviour Mitigation ☒ No visible blank walls ☒ No visible recesses ☒ no visible waste containers ☒ Clean views ☒ Behaviour Direction (sheep effect) ☐ other:	Omnipresence: ☒ Design cohesion ☒ Anonymity ☒ 'highly' management (no effect broken window) ☒ Controlled vegetation Omniscience: ☒ CCTV (presence + signs) ☒ Light (uniformity/all cover) ☒ Panoptical (No blind spots) ☒ Passive surveillance

publicness

Simple Metric:	Right to the Polis:	Aesthetic/ Subjective:	Resume:
Access with neutrality ☐ -1 ☒ 0 ☐ 1 Behaviour Diversity ☒ -1 ☐ 0 ☐ 1 **Uses:** ☒ civility ☐ animation ☒ directed ☐ resilient ☐ presence of counter uses ☒ consume oriented	☐ Ability to occupy ☐ Protest allowance ☐ Happiness (Drinking) ☐ Agency (public decisions on control) ☐ Interest (public benefit directs decision manking) ☐ Intersubjectivity (space facilitate diversity encounter)	☐ Authenticity (place-bounded culture) ☐ Independent actions (not progr.) ☐ Inter-subjectivity ☐ encounter ☐ exchange ☐ difference ☒ controlled ☐ loose ☒ tomb ☐ alive ☒ clean ☐ cosy ☒ dress coded ☐ mix ☒ consumption ☐ use	☒ Tangled Orbits (codes) ☐ 1 ☐ 2 ☒ 3 ☒ Velvet Ground (Physic) ☐ 1 ☐ 2 ☒ 3 ☒ Repeated compulsion for space consumption (Content) ☐ 1 ☐ 2 ☒ 3 PRI ☐ 1 ☐ 2 ☒ 3 PPS ☐ 1 ☐ 2 ☒ 3

Personal Observation
Walking down the street, the 1 New Street Square development is easily perceived. Predominantly grey, the cluster of buildings has gaps in its corners, like spatial fissures that create a passage through the complex. On the other side, in a calmer street, the boundaries snakes the limits of public/private space, and as a lasso accommodate one more bench, in which a working class figure is sited. (TO) The pavement has straight lines of shade change, like the immaterial barriers in the art of Marcius Galan. A small totem with chrome maps, information and a CCTV drawing marks the entrance of the urban fissure. Everybody is wearing suits, some passing by in a hurry. Two figures in black stand in the front of different buildings. As I walk around, I imagine if one of them is watching me, so pictures are taken hidden. (RCSC) Entering the square, many people are sited in cafes, but only one person is sited on the square benches. In Land Securities website own words, the place is thus outlined: "There has been a real sense of magic around New Street Square since it opened, with the dramatically expanded public realm being used for everything from live sports screenings to local food markets"

Photos

Graphics

Fig. 2.5 Example of a field investigation form. (Source: the author)

instance) and becomes a process of reifying the space of everyday life as a whole. Although this process presents itself as 'natural', it has been supported not only by the new set of legislation and the massive strategies of 'regeneration' but also by a good deal of spectacle, rhetoric, and law enforcement.

As we observed, this process transforms users into 'framed free citizens' through the production of three new dimensions of urban experience: (a) tangled orbits; (b) velvet grounds; and (c) spatially driven consumerism. Cosy and kind, domestic and comfortable, these dimensions form a type of bird-nest prison, which differ from the former means of panoptical control and operate through disorientation and refuge rather than discipline and punishment.

Tangled Orbits

The reification of fractal social relations creates tangled orbits of social control. This is achieved through: a constant shift in the rules for the use of spaces; a fragmentation of experience, where users cannot fully master space; a disorientation, and a form of remote control; the production of 'invisible cages' by adding widespread techniques of surveillance associated with objects designed to induce behaviour.

In contrast to the apparent accessibility of these spaces, their uses are controlled and managed by different undemocratic entities. This allows the management of a spectacle of beauty and order, while at the same time delivering a feeling of comfort and efficiency for those who join the rules. Nevertheless, similar to the evasion of surroundings in the dark room of a cinema, as Benjamin (2008) argues for the creation of distracted appropriation, this new urban experience cancels the overstimulation so characteristic of modern metropolitan spaces in order to control and guide experience.

To live in cities, we used to confront contradictory experiences that revealed our incompleteness and the cognitive dissonance of our senses, thus enriching experience (Sennett 2002, p. 371). If Simmel (1950 [originally published in 1903]) once characterized the metropolis as the space where individuals could feel free of the 'orbit' of control of small communities and thus only accounting for 'inner laws', this new fragmentation of public spaces creates a field of entangled orbits, obliging user to 'anaesthetize' their behaviour to adapt to constantly changing rules. Thus, this new 'blurred' boundaries of the cities are the naturalization of 'outer laws' inside the individual.

Velvet Grounds

In the London fieldwork, we saw how a new aesthetic reproduces estrangement disguised as the natural order of things. This is achieved by controlling all the aspects of the environment, so they become cosy and hospitable. This comfort anaesthetizes the urban experience, and citizens are domesticated, enforced by the elimination of politics and dissent from these spaces (which are clearly forbidden). Subjects are framed target profiles, and an automation of the citizen is imposed.

Richard Sennett (2002) challenges the idea of a pacification of the city. For him, this notion of comfort is a social construction, which conceives citizens as passive elements, rather than in a hard gained political arrangement: 'Intense civic bonds arouse from the very play of displacement [...] Yet the ancient city was itself not like a monument to stability' (Sennett 2002, p. 371).

While the embrace of the concept of domestication sounds 'delightful', it is self-evident that it also has patronizing connotations of control and hierarchy. Rather than domesticating, this taming (to 'make less powerful and easier to control', according to the *Oxford Dictionary*) is actually a form of dressage (an 'art [...] that develops obedience, flexibility and balance', according to the *Oxford Dictionary*).

Therefore, these new public spaces are not neutral things. They create homogeneity and comfort in order to frame experience and manage the passivity of object-subjects, so they can become full potential consumers. It is in this sense that these strategies transform the city's spaces into a 'velvet ground' created by comfortable, shining and blurred spaces. Anaesthetized by the constant vibration of 'urban life', the wandering citizen represses his innermost values by experiencing the place only by playing the game of repeated consumption.

Spatially Driven Consumerism

In these new spaces, citizens become instruments and objects of a game of driven spatial consumption. This is achieved by carefully orienting all elements of design to acts of consumption. The whole environment becomes a simulated spectacle, where consumers are themselves appropriated by distraction that turns the whole environment into a fetish of commodity. This is achieved by a careful management of space and anxiety, so citizens enter into a compulsive repetition of consumption to regain the lost appropriation of space by means of a game experience.

Sennett (2002) argues that the urge to experience comfort in cities is a dangerous game because too much protection from 'stimuli' can create a 'sickness of lack'. He argues that pleasure can only emerge as a distension of unpleasurable experiences and that we need to confront ourselves with the 'reality principle' to acknowledge the degree to which our minds are playing with our senses of pleasure and unpleasure.

This idea is founded in the Freudian text *Beyond the Pleasure Principle* (Freud 1955 [originally published in 1920]), in which it is argued that although unpleasure and pleasure are important elements with which to understand mental life, it is the process of the 'repression of desires' that is the key principle. As Freud (1955, p. 11) asserted, 'Most of the unpleasure that we experience is perceptual unpleasure' because our instincts and desires are repressed inside our unconscious by our social experience. This perceptual unpleasure is formed, in such cases, by the expectations of unpleasure and danger, or by the pressure of unsatisfied instincts.

Therefore, a conflict emerges between the repressed unconscious and the actual experience, creating a compulsion to repeat the manifestations of that repressed desire. Those manifestations can emerge as painful experiences or as accepted social games. Freud explains this condition by describing a game played by a child. The child repeatedly throws a toy behind the bed to enjoy reencountering this toy when he finds it. The child achieves, by incessantly repeating the disappearance of an object and its rediscovery, a form of revenge and the feeling of mastery over his destiny. Therefore, the child regains control over 'the passivity of the experience' only through 'the activity of the game' (Freud 1955, p. 17).

Latham and Koch (2013, p. 6), with the story of a 'piano in a square', inadvertently described how the passivity of the urban experience is created by such games. In a regenerated London square, a man saw a piano in the street and asked the waiter if he could sit on the chair, only to be shown a sign reading, 'Play Me. I'm Yours!' After playing music and being applauded by the surrounding customers, who kept asking him for 'one more', the man left. The passivity of his experience in the velvet architecture of London was joyfully overcome through a planned 'game activity' reified into the form of a well-placed piano.

Putting Together the Subtle Game of Experience Orchestration

In the big picture or urban experience, this is a very subtle game. In contemporary entangled spaces, the anxiety of the experience (generated by the multiple reifications of social rules in such spaces) is swamped by the specific game of consumption. The feeling of belonging is carefully orchestrated so that one only feels suited to the space as one engages in the game of consumption. Tables and chairs, as well as views and pianos, are carefully orchestrated by business intentions. Thus, the former blasé 'abstract distance' is transformed into domesticated 'consumer closeness'. This is the repeated game of consumption, a game of incessantly repeating the disappearance of citizenship and its reencounter in the form of commodity.

In the first phase of fieldwork, we saw how the majority of the new public spaces are being transformed into privatized public spaces, and that transformation has a large impact on the public realm. This process is a dialectical arrangement of economic, political, and social doctrines, which creates new codes of interpersonal behaviour in the city. In practice, we learned three things about London's enclosure: fractal social relations are *reifying* tangled orbits of control that disorient users into passivity; spectacle and comfort are the *thingification* of an (an)aesthetic velvet ground, seductive enough to tame behaviour in a bird's nest prison; and the managing of a compulsive anxiety into a game experience *objectifies* citizens into driven consumers (synthetized in the Fig. 2.6). The diagram visually represents the three dimensions of enclosed public spaces, illustrating how urban experience is reshaped into a controlled and oriented spectacle.

Tangled orbits are the result of a fragmentation of rules and experience that renders users passive because they are always estranged from the place, disoriented because not fully aware of what they can do. Feeling at home but not empowered, reinforced by a constant feeling of surveillance. Barriers control where to stay, bollards direct flows, subtle lines in the pavement demark where rules apply, hostile

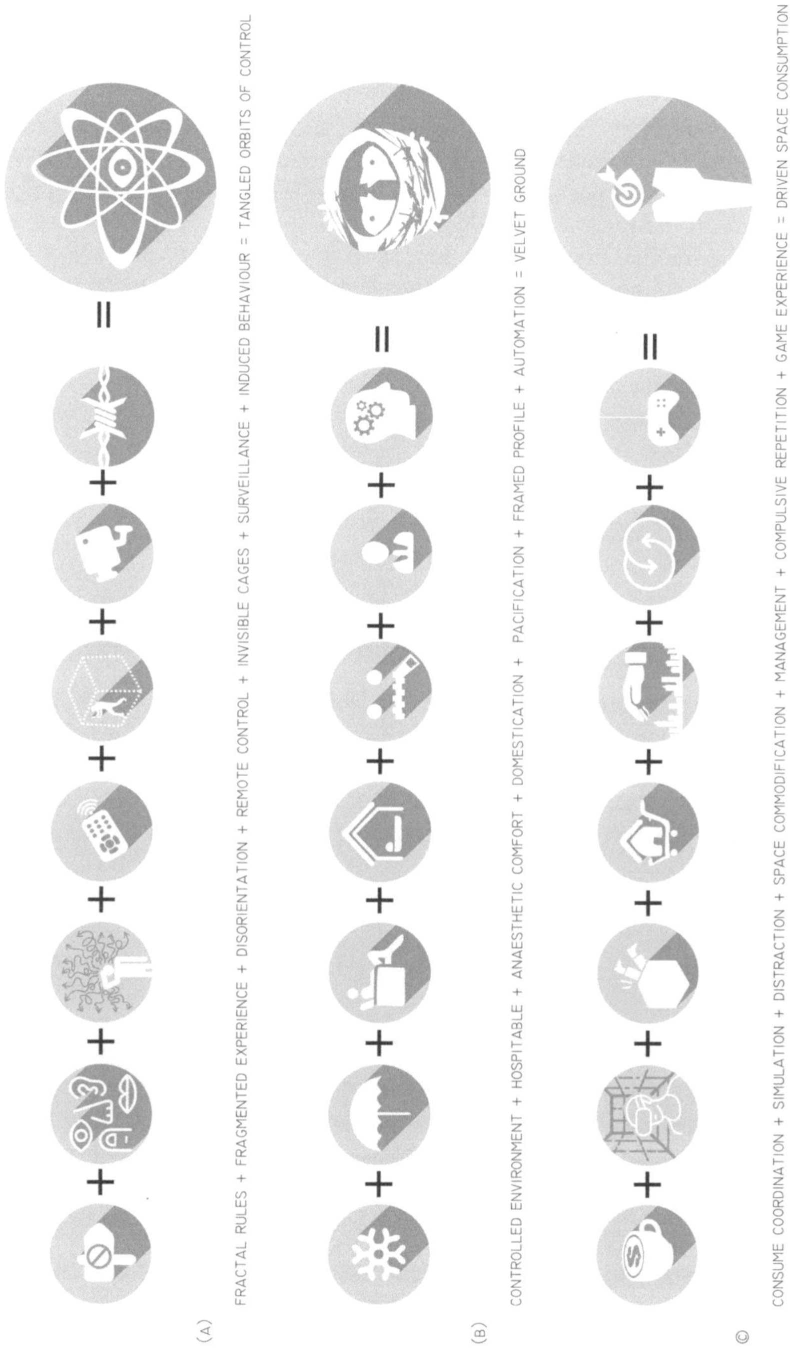

Fig. 2.6 Summary of London's enclosure. (Source: the author)

architectural devices deter skaters and homeless people, furniture and shadows induce actions (paths to go, where to sit, limits of access, and so on), controlling behaviour remotely. The result is an environment where individuals internalize external laws, creating invisible cages that maintains the illusion of accessibility. *Velvet grounds* are produced when aesthetic comfort pacifies urban life. Through controlled environments, framed profiles, and domestication, public spaces become 'bird-nest prisons'—inviting yet restrictive. This structuration eliminates political engagement, fostering mechanical behaviour in a seemingly hospitable setting. Finally, *spatially driven consumerism* highlights how anxiety and urban spectacle merge into a game-like experience. A repetition of coordination, simulation, gaming, distraction, commodification, and management depict how citizens engage with space primarily through consumption. This orchestrated game replaces civic participation with commercial interactions, reinforcing a cycle of passive belonging. This diagram encapsulates how privatized public spaces subtly control and shape urban interactions, turning individuals into 'natural' consumers.

2.3 Against 'Taking Things Easy'

Moving forward, a critical analysis of the arguments in favour of such changes will be helpful to trace back the acceptance of things as natural and to trace back its ethics and veiled elements of control.

For De Magalhães, there is no reason to be 'overpessimistic' about the privatization of public spaces, and pessimistic views on this process come from a frozen, idealized concept of public space (see De Magalhães 2010). He argues that it is just a matter of recognizing the 'natural' necessity of current developments, as the bankruptcy of the welfare state is an unquestionable fact (De Magalhães 2010, p. 560). For De Magalhães, the only possible action is to technically understand the process and, by so doing, improve it. For him, the ongoing process can be described as a simple 'contracting out' of the public character of cities' spaces, and the heart of the matter is knowing how to make good contracts to balance rights and duties. He argues that this is not a history of 'corporate take-over' or 'exclusion' but rather a 'complex redistribution of roles' and a 'domestication' of public spaces.

Similarly, Latham and Koch (2013) argue that the ongoing process of transforming public space is the 'hard work of domestication'. They contrast their use of the concept of domestication with its former critical use (see Zukin 2010). For them, the spaces in question have never been 'wild' because they are artificial products. In this sense, it has not been tamed because it was tamed from the beginning. In these terms, by ignoring dissent, social conflict, economics, politics, and history, De Magalhães, Lathan, and Koch can understand the domestication of urban space as an expansion of 'homemaking' by the blurring of differences between private and public. This provides many points to apply to our analysis.

The first point is the supposed neutrality of the process of 'contracting out'. This assumption disregards both the condition in which these contracts are made, and the positions of power of the subjects that are engaged in the negotiations. In this way, the rhetoric of neutrality acts in the reproduction of social relations in two ways. First, local authorities suffer different pressures, from campaign funding to budget austerity, which might force a choice for abdication. They might even not be aware of other alternatives and be influenced by 'trendy' experiences. Secondly, the social actors involved in the contract are informed differently about the elements and the rights that are been sold (most of the users will never know what happened in this negotiation), while developers are usually aware of the elements at play, as these opportunities emerge from systematic efforts on their behalf.

The citizens, object-subject of the interventions, are in turn submitted to techniques of disengagement and propaganda. Therefore, the proposal that it is simply a matter of knowing how to 'technically control' the process is an academic's delusion. It ignores the political struggles of interest involved. It also ignores the power that emerges from economic and cultural capital: both influencing decisions and producing consensus by financed design and research (we will discuss examples of that in Chap. 4).

In addition, the lived space is fragmented, into a profusion of rules (sometimes formalized in these contracts, sometimes not), transforming the public space into a tangled orbit. Thus, these rules change from place to place, square to square, street to street, and even from a moment to another. The regular citizen, walking through the city, experiences a constant metamorphosis of the city's character, becoming persistently more or less free, more or less familiar, more or less active, more or less detached. For those in the streets, the feeling is a constant shining, blurred and soft, estrangement. The way out of the anxiety induced is managed, so it is found only through consumption. This is one of the ways that the experience of citizenship in the city is instrumentalized.

Furthermore, the complex fragmentation of rules can enable the emergence of malicious domestications (dressage) without any social control over it. Additionally, the blurring of boundaries makes awareness of limits difficult, so individuals become atoms (a small cog), that cannot master the overall picture and the means of space production. Space (the whole) is turned into a machine of reproduction. This condition prevents the emergence of significant differences, as it enforces passivity and eliminates political dissent. Dissent, as the basis of politics (see Rancière 2010), is one of the first targets of these many 'contracted out' rules, usually prohibiting any form of protest and even leafleting. This enforcement of a de-politicization of public spaces—evident during the Occupy movements—is a fundamental shift in the character of public spaces.

The second point is the naturalization of domestication and the formulation of a narrative that fundaments an ethical linear causality: *domestication = cosy + good = tamed + unconditional = citizens' dressage is fine*. Although the human condition is artificial before even the emergence of the *homo sapiens* species,[2] there is a fundamental ethical problem with 'domestication' of citizens as it blurs the difference between 'being a subject of' and 'being subjected to'.

[2]As anthropological archaeologist Timothy Taylor (2010, pp. 160 and 145) argues, 'things rule us' in their 'slippery and redefinable associations', and if they 'appear to have no ideology', 'in the practice that then emerges, everything may change, and your culture and identity may disappear'.

In that proposition, architecture becomes the social engineering of a passive and homogeneous subjectivity, as is evident in Latham and Koch's enumeration of the elements to create domestication:

> **'Foundations'**: the reimagination of the place, which is not a tight script, but is enforced by social and legal legislation (such as 'permits, licensing, management contracts, leasing agreements, service provisions'). 'Important here is the idea that foundations are not only about the material. They are also about a kind of ethos – a way of thinking about the kind of public space that might be put together.' [...] **Furnishings**: [..] 'Engagement with a range of material objects [...] we can outline a process of domestication here consisting of several general features. First, there is the construction of regular, predictable routines associated with the use of these furnishings. Secondly, there is the establishment of roles for taking responsibility for the space and its furnishings. Thirdly, [...] practical activity can generate collective affects'. [..] **Invitations**: [..] 'The hard work of domesticating a public space – creating a new "patterned ground' [...] norms, conventions, expectations, appropriate forms of behaviour and so forth. [...] making a regular habit [...] the routines already described did not magically happen on their own. People needed to be invited to make themselves [...] **Accommodation**: [..] 'Closely following this example, a fourth element in the domestication of public space is accommodation: the process of adapting or adjusting to others in order to get on with living.' (Latham and Koch 2013, pp. 15–18).

Despite the 'delightfulness' provided to those willing to obey, there is a problematic effect of this line of thought. As Simmel (1950) asserted, the foundation of modern rational freedom in the metropolis is based on the experiences of abstraction, distance, and difference that are made possible by the big city. In opposition to this modern space, the feudal city was formed precisely by a domesticated space. A space that kept people locked in the condition they were born, without way out. This condition was fully experienced through emotional relations to the place, and mediated by the 'heart' and tradition, with the sensation of a cyclical and eternal stability that can only be felt at home. Therefore, the proposition of transforming the modern city into a home is, by logical conclusion, regression from a long-fought battle for freedom in the modern metropolis. A freedom based on the abstract distance and solitude among the multitude, which is so important to creating the uncomfortable encounter with the Other. As Simmel (1950, p. 418) argued: 'For here as elsewhere it is by no means necessary that the freedom of man reflect itself in his emotional life only as comfort'.

The third point is the false dilemma of either we accept 'things are as they are' or we fall into 'big state fascism'. The critique is not about reaffirming the state's lost control of space. As Zukin (2010, p. 30) asserts, in the privatization of public spaces and regeneration schemes, the state plays a strong role in guaranteeing that big business gains control over the city. As she demonstrated in the case of the World Trade Centre in New York, in these schemes, state control can become more repressive, more ideological and less representative than in former types of

---

Taylor (2010) argues that things are changing humans not only in modern times but also since the beginning of our species because evidence of the first stone tool dates back 190,000 years before the evidence of the first specimen of the genus Homo. These tools allowed a diet rich in protein and, in turn, the evolution of big brains in first place, and even the control of fire has reduced our body hair. Therefore, for Taylor, there are no natural humans: we are 'artificial apes'.

intervention (Zukin 2010, pp. 150–158). Finally, one last point is straightforward: it is naive to consider any 'thing' as 'natural', as it is to consider any 'fact' as 'unquestionable' although these are the basis of their arguments.

Beyond the fact of progressively separating the citizen from the means of freely appropriating the place and at the same pace reifying behaviours, these new dimensions have mechanisms that hide the place commodification and create a sense of a 'naturally artificial' social order. In this sense, to unveil how the reification of contemporary spaces is changing social relations and citizens, we will need to develop a critical and dialectical account between physical design and economic, political, and social processes. If architectural things are social codes and interpersonal rules engraved in space, then these common places are not only a matter of built barriers, choices of cosy benches, or elegant fancy materials but also a matter of objectifying social relations reproducing and reinforcing citizenship roles. We will further develop this conceptualization in Chap. 3.

Chapter 3
For a Theory of Urban Reification: Designing Urban Things and its Fantasies

Abstract This chapter sets out the basis for a theory of reification (the objectification of social relations), based on concrete examples presented in the previous chapter. It sets a cognitive map of how ideology and desire are used to frame urban experiences. In addition, it traces how architecture produces narratives (fantasies) that, in their turn, reproduce ideological and subjective senses (the way people see social reality). Finally, it develops an alternative route to enable awareness of the spectacular character of these neoliberal spaces.

Keywords Conceptualizing reification · Things · Urban reproduction · Social relations · Spatial instrumentalization · Neoliberalism · Consciousness · Subjectivities · Desire · Fantasies

In the previous chapter, we aimed to arrive at a first-hand account of the relation between the production of urban 'things' and the reproduction of social relations. For that it draws upon a literature review on the role of public spaces and a fieldwork investigation on privatized urban spaces in London. The analysis shed light on the relation between the design of public spaces and the reproduction of subjectivities. That also aimed to shed light on the contrasts and continuities of social relations mediated by contemporary metropolizes. Thus, what has been observed by scholars about the modern metropolis and the new aspects observed in the field can be compared.

Although the relationship between the metropolis and the reproduction of social relations have changed in different historical and cultural contexts, and cities have played different roles in the formation of modern citizenship, in these new common spaces, old hierarchical social relations are intensified under the guise of new appearances. Therefore, rather than repetition, this process reproduces social relations through new mechanisms objectified in space. These new appearance of 'things' in cities creates a specific disciplining of subjectivities, and we unveiled the aspects of this new urban aesthetics: Velvet Ground, Tangled Orbits, and Spatially Driven Consumerism. That launches the basis for formulating in this chapter a theory of urban reification.

C. V. De Lima Amaral, *The Urban Reproduction of Subjectivities*, SpringerBriefs in Geography, https://doi.org/10.1007/978-3-032-01465-8_3

What follows is a critical theory of urban reification, aimed at unveiling the protocols, desires, and fantasies embedded in spatial production. This allows us to reframe public space not as a passive stage, but as an active instrument in the reproduction of social order and, potentially, as a battleground for its transformation. Later, in Chaps. 4 and 5, we will explore the methods applied by urban designers and stakeholders to reproduce this reification of social relations—namely, fetish—so in Chaps. 6 and 7 we can investigate how critical practices are still trapped in the subjectivities objectified by disciplinary apparatuses. For now, we aim to investigate how urban 'things' can be something more than neutral objects—namely, social relations objectified in space acting as if they were natural and unconditional parts of reality.

3.1 Excavating the Social Foundations of Urban Enclosure

The aim here is to uncover the differences and continuities between these new privatized spaces and the previous ones that supported its emergence. In this way, we can trace back the roots of contemporary contradictions and develop a theory of how urban things reproduces social relations—i.e. how reification operates. Urban spaces are not mere physical settings but complex arenas where power, desire, and social relations are constantly produced and contested. This section sets out to excavate the social foundations of urban enclosure, interpreting the transformation of public space through the lens of reification and primitive accumulation. As cities increasingly adopt privatized, securitized, and corporately managed environments, traditional democratic spaces are being subtly dismantled. Arguably, this process started with developments like Canary Wharf, and it is not a matter just about ownership, it's about the material and symbolic restructuring of the urban condition.

New Common Places in Town

At the end of 1980s, we saw the birth of a new kind of space as a result of political and economic reconfiguration. The metropolis in centralized economies faced a process of deindustrialization, which, with the increasing flexibility of work relations and the pressure on the welfare state, reframed the mass production economy into the so-called flexible economy (Harvey 1990). For this new neoliberal economy, a new form of space was produced. The paradigmatic example was Canary Wharf. Developed in an old industrial port area, a vast new financial centre was designed from scratch, where the streets and squares were no longer 'adopted' by the weakening welfare state but were controlled by another entity: the corporation.

Zukin (2010, p. 222) states that this is a long process that took place in the period after the great wars, creating what she called the 'corporate city'. This city has been produced by specific arrangements of: private capital investment, State policies and rules, design, media image construction, and direction of consumer tastes (Zukin 2010, p. 30). This new configuration of social space (capital, state, media) excluded civil society from the formula. Not surprisingly, the collapse of public spaces is accompanied by the collapse of democracy itself, as empirical studies demonstrate that political decisions are defined by corporation interests and are no longer influenced by the public opinion (Gilens and Page 2014).

In the book *Ground Control*, Anna Minton (2009) analyses the emergence of privatized public spaces, arguing that Canary Wharf became a model for almost all recent developments in England (Minton 2009, pp. 3–14). Although maintaining the appearance of public spaces, these streets are not democratically accountable. Such spaces are privately managed and are based on specific strategies and sets of rules. The result is high levels of management, stratification, big chain stores, and corporate control over public security to avoid any disruption of capitalist activities. Besides maintaining the appearance of public spaces, these streets are not adopted by the local government.

For Smith and Low (2006, pp. 14–16), as we watch the collapse of basic rights, public space emerges in the centre of the contemporary debate on democracy, as the production of space is a central strategy for the implementation of neoliberal concepts. For Smith and Low, this is the reason why political movements are always attached to places (the archetypical mass in a public space). As we will see, changes in the production and control of public spaces will also have implications on the political potential of those places. Zukin (2010, p. 145) argues that schemes of private city space management are a new form oligarchy. For instance, the votes and decisions in Business Improvement Districts are balanced by the size of ownership in the local arrangement, and usually the vote is proportional to the size of the property. Based on the argument that the State does not have funds in a moment of economic crises; therefore, those who have capital should rule. For instance, in this equation homeless and unemployed people simply do not exist. Smith and Low (2006, p. 15) consider this is the formation of a new apartheid, excluding the poor, the homeless, and immigrants out of the management of city spaces.

Furthermore, local authorities in the UK now expect buildings and streets to be designed according to the guidance of 'Security by Design' manuals (Minton 2009, pp. 61–82). These manuals establish a series of design measures that incorporate implicit strategies into physical objects. Thus, space is filled with intentional artifices that are designed strategically to avoid conflict, control use, orient behaviour, and guarantee business-friendly environments. Therefore, both Appleton (2014) and Zukin (2010) emphasize that laws are increasingly a mixture of authorities' wish to control society and business' interests framing decisions in the management of social spaces.

Besides these laws and spatial tools like 'Secured by Design' aimed at controlling the physical aspects of spaces, other legal tools—such as Public Space Protection Orders and protest-targeting statutes—reinforce how these spaces can be used. Using the rhetoric of safety and public order those laws serves a dual function: justifying urban redevelopment and criminalizing political dissent. In this double movement, the discourse of security displaces democratic concerns about space.

Moreover, Appleton (2014) argues that these laws give authorities an almost free hand to control who can do what in public spaces, with complete flexibility in rule design. In this sense, there is no previous definition of what is prohibited, nor any previous definition of the scale of the sentence, thus holding people in a state of permanent self-vigilance. Along with many mechanisms to assure dispersal powers to the police, these new laws create mechanisms in which specific punishments can address specific individuals and behaviours, depending on the understanding of a court (Minton 2009, pp. 132–178). Thus, the modern idea that public space is a place in which people can manifest their freedom has been significantly diminished.

To reinforce control of behaviour in urban places, the Anti-Social Behaviour Act (2003) introduced a legal framework that has evolved over the years. Although Anti-Social Behaviour Orders (ASBOs) were abolished in 2014, they were replaced by new laws, continuing the trend towards spatial governance through targeted behavioural sanctions. For Boffey (2023), these new laws collectively erode protest rights and urban space freedoms. The Police, Crime, Sentencing, and Courts Act 2022 broadened police powers to impose conditions on protests. The Public Order Act 2023 introduced new offences for peaceful manifestation that causes 'more than minor' disruptions. And the related Serious Disruption Prevention Orders increased the power of police to restrict and control activists even before protests takes place. Boffey warns these measures risk chilling democratic dissent by setting vague grounds and essentially reframing protest as a threat to be managed rather than a core civic right.

Gayle (2025) reports that Yasmine Ahmed of Human Rights Watch deeply criticizes Labour's continuation of protest laws. She argues that these laws have drastically lowered the threshold for police intervention, increased arrests, and harshened penalties. These measures demobilized citizens participation, marking it a dangerous precedent for democratic rights worldwide. Michel Forst (2024), UN Special Rapporteur on Environmental Defenders under the Aarhus Convention, have also strongly criticized these developments as authoritarian in scope. Akiko Hart (2025), underscores the disturbing contradiction that even those who voted against anti-protest laws, once in power did defend them in court, revealing how even supposedly progressive governments may be seduced by authoritarian powers. Furthermore, the UK Home Secretary threatened to use the Terrorism Act on Palestine protestors, with possible imprisonment of up to 14 years (Siddique and Abdul 2025). These trends exemplify the shrinking space for dissent in urban spaces.

In sum, the transformation of London's public spaces since the late 1980s is more than a shift in urban design. It marks a profound restructuring of social and political life. What began as corporate-led redevelopment has evolved into a broader regime of spatial control, where security rhetoric, legal instruments, and privatized planning converge to suppress dissent and displace democratic participation. These new common places, while appearing open and accessible, operate through mechanisms of exclusion, surveillance, and dispossession—normalizing a cityscape where market logic suppresses civic rights.

Background Noise

Specifically in London, the struggle for the control of public spaces and streets was thc outcome of a long dispute that happened towards the end of nineteenth century (Minton 2009, pp. 19–21). By this time, major squares of the city were enclosed, guarded by private security and surrounded by sentry boxes. After a conflict with guards resulted in a murder, social unrest and the involvement of the printed media resulted in two major parliamentary inquiries. These inquiries resulted in streets becoming 'adopted' by the local authorities,[1] which were gaining power by this time. In this sense, the rise of the public space coincided with and represented the construction of local democracy in London.

Many intellectuals have observed the connection between the space of the modern metropolis and construction of social freedoms (Simmel 1950; Lefebvre 1991a; Berman 1990; Baudelaire 1995; Sennett 2002, inter alia). Marshal Berman (1990) examines how different cities in the nineteenth century transformed the citizens' social perception. Contrasting the boulevard's experience in Paris with the Prospect's experience in Saint Petersburg, Berman demonstrate the intimate relation between the social awareness and the city forms. Those spaces transformed the urban experience into a social experience that contrasted collective consciousness with the constraints of the human condition and, so, into a direct clash between different classes. For Berman, that is the reason why the modern metropolis was the fundamental trigger of the most important social movements of that century.

Charles Baudelaire is perhaps the most cited author on the account of the modern construction of freedom through the experience of the city. In his literature, he was able to express how the modern metropolis became an instrument to create an ephemeral and free spirit in the modern man (cf. Baudelaire 1996). For Baudelaire (1995), the modern man is like the modern painter, who can design his own life, as freely as an artist design his paintings. This freedom was enabled by the sudden transformation of the small, closed, and stable traditional space, into an open and dynamic space filled with a multitude of diverse people. This atmosphere was founded in the city itself and was experienced by the *flaneur* attitude: a modern man that would engage in the news and movement of the city, connecting freely with the street's 'family of eyes', and swimming in the streams of passersby (Berman 1990, pp. 148–154).

In this sense, Sharon Zukin (2010, p. 129) argues that public space has had an important role in the democratization process. For instance, even before the French Revolution clashed against the previous aristocratic privileges, the mixture that was promoted by the markets' space was creating the basis for modern democracy and paving the way for the city space to be open for all. Moreover, Zukin highlights that

[1] Property of streets and squares in the UK remained largely private. Thus, when the Occupy London movement began in 2011, the only place to occupy was the square in front of St Paul's Cathedral. Since no central London space was truly public, only the Church was willing to tolerate the protestors.

London and New York created public spaces in the nineteenth century, such as libraries, parks, and museums, even before the right to vote was available to everyone. Rather than just an effect or image, public spaces were also generating a feeling or desire for democracy and are, therefore, politics in praxis.

Moreover, Georg Simmel (1950), from a different perspective, analysed how the experience of the metropolis changed modern men's consciousness and, therefore, social relations. For Simmel, the modern city created an intensification of 'nervous stimulation', a continuous confrontation with the unexpected, a highly impersonal relationship with others, an abstract mode of mediation (money) and, therefore, the necessity of fast reactions. That situation led to the unconditional use or rationality. In other words, by these new features of the metropolis, the modern man is ripped apart from the old 'orbit' of the feudal village. This idea of 'orbit' will be latter important for understanding how new privatized public spaces are creating a tangled orbits of control. Therefore, for Simmel the modern metropolis experience generates freedom for individuals by the transcending of its own displacement, thus individuals pass to only accounting to its 'inner laws'.

Nevertheless, David Harvey (1990) and Fredric Jameson (1991), among others, have discussed how changes in the mode of production have transformed the way spaces are produced with the intention of reshaping the way society operates. They argue that the move towards a post-Fordist mode of production is articulated with many aspects of post-modern urbanism and spectacular architecture. This aims to change the very nature of work, society, control, and space, arguably to adapt it to a new condition formed by contemporary logistics, automation of production, and cybernetics (Negri and Hardt 2000; Srnicek 2013).

Michel Foucault (1980) and Gilles Deleuze (1992) argued that, with modernization, capitalism moved from a centralized society, controlled by despotic power, with hierarchy and direct modes of punishment and control, towards a disciplinary society, organized by a strong bureaucracy, where the decentralization of control was enabled trough rigid set of laws and education. Henri Lefebvre (1971a) called that society 'The Bureaucratic Society of Controlled Consumption'. Nevertheless, in one of his later works, Deleuze (1992), asserts that society was moving towards a new form of social organization, where the economy would be organized by cybernetic machines and computers, with a new kind of social control being born, which is much more diffuse and implicit. Arguably, these new neoliberal spaces are a part of that stage.

In the same direction as Deleuze, Alexader Galloway (2004) argues against arguments such as Castells' (2002), which considers the network implicitly free. He argues that today control is established in the same manner as the *protocol* controls the internet. Protocols are the system that rules exchanging data between different machines. Because software is made of algorithms, he argues that control have become much softer. Therefore, he compares this softness of control with the kind of control one can feel in the highway:

> To help understand the concept of computer protocols, consider the analogy of the highway system. Many different combinations of roads are available to a person driving from point A to point B. However, *en route* one is compelled to stop at red lights, stay between the

> white lines, follow a reasonably direct path, and so on. These conventional rules that govern the set of possible behavior patterns within a heterogeneous system are what computer scientists call protocol. Thus, protocol is a technique for achieving voluntary regulation within a contingent environment. (Galloway 2004, p. 7)

Therefore, Galloway uses the architecture of the internet to generate powerful insights into contemporary structures of control—particularly through the dual nature of internet protocols like TCP/IP and DNS. These systems operate through two abstract mechanisms: one enables a kind of formatted freedom (similar to how highways direct but allow movement), and the other enforces hierarchical control—like the centralized authority that, as Galloway (2004, p. 10) notes, can erase all mention of 'China' from a country's internet. He concludes that what the commodity was for Marx—the object through which capitalism mystified its social relations—the protocol is for our time: a technical form through which power and control are now exercised invisibly.

Just as this dual structure is omnipresent on the digital world, it also shapes contemporary urban spaces: public areas that are formatted by the State and privatized public spaces controlled by corporate interests. The result is a spatial logic in which individuals are free to navigate—but only within predefined rules that enforce order and prescribe behaviour. What is new is that these spatial protocols now govern a wide range of urban experiences, from crowd control and tourism flows to property rights and driven consumerism.

Douglas Spencer (2011) has studied how contemporary architecture is producing spaces with complex forms, smoothness, folding, and responsive to difference, but, at the same time, conceiving the buildings in order to achieve a new kind of smooth and invisible form of control.[2] They transform former critical concepts from a form of 'critique' to a form of acceptance and 'valorization', manipulating the current traits of contemporary neoliberalism into a new kind of 'politics of affect', an a-political politics (Spencer 2011, p. 18). Therefore, terms once used to build a critical and negative analysis (such as nomadism, differential, deterritorialization, micro-politics) are now used to achieve its furthermost neoliberal potential. Thus, these seemly fluid, free, and liquefied spaces are, intrinsically, controlled by hidden artifices.

Nevertheless, Alex Williams (2013a) argues that the idea of 'everything being capitalism' is either fallacious or a misunderstanding of how systems' works. He argues that the entity called 'capitalism' cannot be omnipresent and omniscient because it is not a 'closed system'. In contrast, it is an assembly of many open systems interacting with each other and fighting for hegemony. He concludes that although Capitalism is totalizing—projecting forms of control to englobe all the others—it is not absolute—fissures and cracks and other spaces remains in the fight.

---

[2] Spencer (2011) analysis a new educational building, for accommodating Ravensbourne's relocation to Greenwich, showing how educational spaces and 'public spaces' are carefully interconnected to make the students engaged with marketing at all times, whereas the public is invited in the building as 'consumers'.

Therefore, it is important to investigate how dissent and alternatives may be produced within—and maybe even by—the capitalist space itself.

Primitive Enclosures

The process of transforming space to introduce new forms of social relations is not new. In Part VIII of *Capital* (1990 [originally published in 1867]), Karl Marx famously analysed the process of privatization of common land in England as an example of a form of 'primitive accumulation'. In that context, he demonstrated how capitalist development in England was preceded, on the one hand, by a legislative process of wealth expropriation that focused on confiscating common land and transferring its ownership into private hands. This created a dispossessed and dependent mass of proletarians. Examples of this process were the 'Enclosure Acts' created by Parliament (Chapter 27). But not only that; on the other hand, new legislation was produced to criminalize the behaviour of the poor through a series of statutes that were created to enforce a culture of wage labour behaviour and to control 'beggars' and 'vagabonds' (Chapter 28, entitled 'Bloody Legislation…'), thus establishing a disciplined and dependent proletariat.

De Angelis (2000) argues that Marx's 'primitive accumulation', as explained in the 'land enclosure' process, was not a process concealed in time and space; rather, it is a continuous process of wealth accumulation that constantly reproduces inequalities based on access to power. In this sense, the 'primitive' has a sense of 'in anticipation', i.e. a primary foundation for capital accumulation that is the basis for the separation of workers from the means of production. For De Angelis, this separation is the grounds of Marx's theory of reification: the transformation of free subjects into objects in the marketplace through the alienation from the instruments of production, thus imposing on them the condition of what once was wage labour and now is consumerism reproduction. Therefore, he argues that 'primitive accumulation could be identified in those social processes or sets of strategies aimed at dismantling institutions that protect societies from the market' (De Angelis 2000), demonstrating how this procedure is constantly re-emerging and even reinforced in the neoliberal context.

In this scenario, Ross Adams (2014) argues that architecture in contemporary society is becoming a technology of subjectivity, functioning as a means of crystallizing specific forms of socialization while pretending to be a-political. For him, architecture has become an 'inferno of the same', admitting no negativity or difference. Furthermore, architecture becomes a meta-architecture, an expansive 'interior' encompassing the whole environment. This meta-architecture is subtle and light, almost immaterial, and its means are transparent and seem to entail no distinguished face, only fascination. Therefore, Adams argues that architecture becomes as vast as the world, an 'act of pure enclosure', which he suggests is the proper phenomenology of empire (as in Negri and Hardt 2000).

As we are about to analyse, London's privatization of public spaces follows a path similar to that of the Enclosure Acts: it is focused on strategies of both 'behaviour control' and 'dispossession'. The process of behaviour control is enforced by strategies of security-by-design, and the public spaces privatization, as well as by the increasingly severe legislations such as the 'Anti-Social Behaviour Orders', The Police, Crime, Sentencing and Courts Act 2022, the Public Order Act 2023 and the Serious Disruption Regulations 2023. These new conditions of legality in the urban production tend to produce a new form of public space. In this sense, just as the very appreciated idyllic landscape of the Highlands in Scotland is the result of the Enclosure Acts, which violently cleared peasants from the land (Marx 1990, Chapter 27), the velvet architecture of London is the result of a new process of enclosure, a re-emerging primitive accumulation: this time in urban spaces and subtle and deceptive; it is a process of urban reification by comfort and disorientation, rather than disciplining and punishment.

3.2 Getting One Thing Straight: Things Are Social Relations

The word 'reification' comes from the Latin *res*, which means thing or object. Hence, reification is also referred to as thingification or objectification. These three terms have been used interchangeably and are central terms in Marxist theory. Although all three terms refer to the same process, here we shall use 'thingification' to emphasize the naturalization of social aspects, 'objectification' to emphasize the instrumentalization of space and subjects, and 'reification' to emphasize the conversion of social relations into things and vice versa.

Reification is a form of 'hypostasis', a fallacy that takes abstractions as if they were a concrete entity (e.g. the map for the territory), but furthermore, reification is a modern social process of imposing historically designed and abstractly produced modern social relations as if they were inevitable, and thus, it is connected to the birth of capitalism in the fifteenth century. Thus, arguably, reification in architecture could be said to start that century, when Alberti (1988, originally published in 1485) first called architecture *De Re Aedificatoria* (literally *the building thing*); and it is a notion that persists through the whole history of modern architecture and still plays a key role in the contemporary ideology of architecture.

For instance, the main theme of the Venice Biennale of 2014 directed by Rem Koolhaas was conceived as a collection of 'fundamentals', which was presented as a series of 'things' (see OMA 2014). These things were understood to be the 'fundamentals of our buildings, used by any architect, anywhere, anytime'. These things were understood to be: floors, walls, ceilings, roofs, doors, windows, façades, balconies, corridors, fireplaces, toilets, stairs, escalators, elevators, and ramps. Here, it is not enough to notice that not a single one of those 'things' was present in the first image at the entrance of the Brazilian pavilion: an indigenous architecture and no floor, no wall, no ceiling, no roof, no door, no window, no façade, no balcony, no corridor, no fireplace, no toilet, no stair, no escalator, no elevator, no ramp, yet, architecture. Rather, the radical step emerges by revealing how Koolhaas' office concretely produced this thingification of architecture.

This reification of architecture was designed upon a reification of other architects. Initially, by conceiving the exhibition 'Monditalia', Koolhaas' office (see OMA 2014) captured other biennales and festivals of the city (such as dance, music, theatre, and film) into his own—thus the whole cultural activity of the city became one single thing. In addition, in one in a series of 'for the first time' in Venice Biennales, Koolhaas' office proposed an overall idea to be addressed by the national pavilions—namely, how the modern movement built the architectural aesthetics of each country. In this way, from the start, Koolhaas' office framed the 'Others' proposals into an inquiry that was his own. Furthermore, this formatting of proposals was reproduced in the part of the exhibition that was supposed to be of his own personal initiative; it was produced in an assembly line of design.

The exhibition was divided into small pieces, each thing designed by a different group of architects. In an interview with one of the architects who participated in the production of the exhibition (who preferred to remain anonymous), it was revealed that his office was approached by one member of Koolhaas' staff and was offered 20,000 euros to design and build one of the elements of the exhibition. Without ever meeting Koolhaas, their first idea was refused for being too radical, too political, and too polemical. A second proposal was made and approved. It remained only 'too polemical'. The budget was too small to cover the costs of the construction of the exhibition piece, let alone the labour time, transportation, daily costs, and lodging. However, the possibility of participating in a famous event, hosted by a famous architect, kept the participants motivated. This partition of the exhibition into an assembly line of design not only reduced architecture to atomized things, but it also reduced architects themselves to objects instrumentalized by one mastering will.

A Sure Thing of All Things

Arguably, the essay 'Reification and the Consciousness of the Proletariat' from 1922, by Georg Lukács (1971), was a benchmark of this debate and influenced many studies that followed. Here, we do not aim to discuss all aspects and implications of the work of Lukács, as our aim is to clarify how the discipline of architecture contributes to the process of estrangement by both reproducing social relations as well as by reifying the subjectivity of architects (or what Lukács called the 'soul').

For this reason, we shall start with Lukács' own 'self-critique' in the *Preface to the New Edition* of 1967 (Lukács 1971, pp. ix–xxxviii) as a means of introducing a radical view of the production of objects (things) and subjects (souls). There, Lukács notes what he considers a 'fundamental and crude error' of his previous work: the matching of reification and alienation. For him, this comes from Hegel's notion that 'the object, the thing exists only as an alienation from self-consciousness, to take it back into the subject would mean the end of objective reality' (Lukács 1971, p.

xxiii). Alternatively, Lukács' self-critique recognizes that objectification is an instrumental device of the mind and is part of both 'true' and 'false' approaches.

> Every externalisation of an object in practice (and hence, too, in work) is an objectification, that every human expression including speech objectifies human thoughts and feelings, then it is clear that we are dealing with a universal mode of commerce between men. [...] And in so far as this is the case, objectification is a neutral phenomenon; the true is as much an objectification as the false, liberation as much as enslavement. (Lukács 1971, xxiv) [Thus he concludes:] For no purposive activity can be carried out in the absence of an image, however crude, of the practical reality involved. Practice can only be a fulfilment and a criterion of theory when it is based on what is held to be a correct reflection of reality (p. xxv).

Lukács identifies alienation as the problem of reification, noting that only when the existence of man is 'subjugated, deformed and crippled can we speak of an objective societal condition of alienation and, as an inexorable consequence, of all the subjective marks of an internal alienation' (Lukács 1971, p. xxiv). In this sense, he was criticizing the following use of his theory, which resulted in a generic critique of any form of rationality, as the ideas in his 'book were converted into fashionable notions' (Lukács 1971, p. xxxviii). In this sense, his self-critique is directed to the possible misuses of his theory. By stressing that there is 'good' and 'bad' objectification, he is condemning the schools that equate any object in society with 'bad', as he suggests is the case with 'messianic negative idealism' (i.e. the impossibility of reaching truth in the absence of the messiah, but in idealistic negation).

As Lukács himself does recognize other achievements of his work, we shall bear in mind the fact that the production of objects and things is a human condition. Thus, we focus on the identification of estrangement: how things become as pure 'facts', and separate subjects from awareness of how they are created, produced, invented, and historically imposed (Lukács 1971, pp. 128, 130, 140). This is the key to identifying how subjects comprehend (and are estranged from) their own positions in the system of production, and how there are always social processes inherent in apparently objective things (Lukács 1971, p. 179).

The fundamental point of Lukács' book was to show how things are not absolute facts. Rather, they are 'produced' in a social and historical process; therefore, things are the result of an 'action', which presupposes a subject who acts and, thus, more fundamentally, 'creates' (Lukács 1971, p. 140). We must acknowledge that we shall have raised ourselves in fact to the position from which reality can be understood as our 'action' (Lukács 1971, p. 145). Thus, the 'action' is the fundamental epistemological key to the process that unites history, social conditions, subjects, and objects.

For Lukács, man is the product of a social milieu, and the social milieu is the product of man; this is the 'epistemological problem of production, in the systematic question of the subject of an "action", of the "creator of a unified reality"' (Lukács 1971, p. 135). Therefore, he points to 'the decisive problem of this line of thought: the problem of the subject of the action, the subject of the genesis' (Lukács 1971, p. 145), which can be found in 'the identical subject-object, the subject of action' (Lukács 1971, p. 149).

In this sense, we aim here to emphasize the relations between architectural things and the production of subjects: how conscious these subjects are about these processes; how the production of things (reification) concretely reproduces social relations (alienation); and how subjects accept these things as 'facts' beyond the action of concrete subjects (estrangement). Thus, instead of accepting or conceiving urban artefacts as a 'thing', an immutable 'absolute', we aim to disentangle in how it simultaneously produces 'things' and 'subjects'.[3]

To explore this, Lukács proposes breaking the immediacy of things and analysing how totality integrates each singular thing in human existence (1971). He describes reification as an inherent reality of society that can only be disrupted by recognizing the contradictions within social development and becoming conscious of their implications (Lukács 1971, p. 197). This approach underscores the dialectic between the part and the whole. As in fractal geometry or in a kaleidoscope, any change in the whole changes the part and vice versa. Therefore, although the part manifests idiosyncratic appearances—which might lead to the error of acknowledging them as individually independent—the individuation of the part is only an actualization of the whole. And here the transindividuality of Simondon (2013) might come in help, bringing Marx and Simondon closer.

In Marx, things are social relations because: they were historically produced, and therefore relate to the context and the concrete labour that produced them; the ideological apparatus (values, knowledge, science, laws, believes, values, and so on, mobilized to see, recognize, and represent things in thought and action) is historically and socially produced; and not only things represent abstractions (as in the case of money), but they also mediate the relations between subjects. Therefore, the accepted 'nature' of things is produced by complex interactions of subjects in a society.

Simondon expands this to the individuation of all things. For him, individuation of things occurs by the production 'mediations', where form and content are mediated by an action (an energetic process, labour). To shape—where (in)formation is to *mise en forme*—is to create a mediation between different orders of magnitude—the part and the whole—upon which a potentiality is actualized. Therefore, things are the shaping of a limit, and even a living individual is a 'theatre of individuations', constantly re-enacting the relations of interior-exterior mediated by these limits and mediations. As a result, bringing Marx and Simondon together enables us to understand how social relations are not simply constitutive of subjects but are also constitutive of all beings, and each little thing.

[3] In this sense, Object Oriented Ontology or Object-Oriented Philosophy simply misses the fact that objects are produced, and thus have a history, a social condition and subjects defining them, and for this reason it is not a critical philosophy.

For a Theory of Urban Reification

As we have seen, Lefebvre (1976) argued that capitalism has managed to avoid crisis by reproducing the relations of production. According to his argument, this reproduction is achieved by destroying traditional spaces and by producing instrumental logic in space. Thus, architecture (and the city) has become the means for trapping society inside the capitalist system, mainly by establishing the fetish of the commodity as a guiding principle of its production. What is a specific logic becomes 'natural' (supposed neutral space). This process could be called spatial reification.

The means of reification have many variations and subtleties of interpretation, having been described in the work of many authors under different names, such as mystification, instrumentalization, phantasmagoria, or protocols (see Marx 1996; Adorno and Horkheimer 1996; Lukács 1971; Galloway 2004). The inauguration of this discussion can be found in an important passage in Marx's Capital:

> The mysterious character of the commodity-form consists therefore simply in the fact that the commodity reflects the social characteristics of men's own labour as objective characteristics of the products of labour themselves, as the socio-natural properties of these things … It is nothing but the definite social relation between men themselves which assumes here, for them, the fantastic form of a relation between things (Marx 1990, pp. 164–165).

Introduced in *Capital* by Marx, it focused on how commodities obscure the social characteristics of human labour by presenting them as intrinsic properties of objects themselves. Unlike Marx's focus on commodity fetishism, our discussion narrows down on urban reification—the transformation of social relations in things in the city. This phenomenon leads individuals to experience the city as an alien force beyond their control. The goal here is not to construct an absolute theory of reification but to enable self-critical engagement with it, following Lukács' melancholy (1971) to perform a constant ethical unveiling of subject-object constructions.

Lukács' approach in 1922 (Lukács 1971, pp. 83–209) has his own ethical commitments: it departs from the understanding of knowledge divided into two main blocks (one bourgeois, the other in the direction of the negative consciousness of the proletariat). Rather than consider it a simplification, we could understand it as a theoretical tool used to clarify points of view and empower an ongoing struggle. Thus, on the one hand, he enables us to understand how the first type of knowledge departs from the 'immediate', and the latter from a 'mediated', point of view. He goes on to analyse what he calls 'bourgeois thought', treating it as a coherent whole only to envision how different philosophical systems are variations of the acceptance of immediate objects as the sources of universal laws. Things become absolute facts.

For Lukács, therefore, that thought builds a pragmatic discourse on truth by departing from those supposed 'facts'. Thus, instrumental rationality and calculability accepts the status quo as an axiom, and the existent objects as natural absolutes. Thus, this way of thinking ignores what he called the 'social processes': the continuous and complex transformations of social things, the historicity of the social relations that had produced those things in the first place, and most important of all, the fact that all knowledge and social existence are produced by humans themselves. In this sense, Lukács argues that bourgeois thought, and practice necessarily mythologize what 'exists' as unconditional and what 'ought to be' as unthinkable (Lukács 1971, pp. 126, 150, 155, 160–163).

Conversely, Lukács critiques theories that elevate mediated knowledge into an absolute, such as Hegel or Kant's separation of reason from the 'thing-in-itself'. This abstraction perpetuates the *status quo* consciousness as immutable (Lukács 1971, p. 150). Lukács warns that accepting a socially produced order as absolute leads to fatalism, wherein social conditions appear as unchangeable natural laws (Lukács 1971, p. 191).

Lukács (1971, p. 203) argues that reality is not a fixed entity but a process of becoming, where things transform dynamically. Recognizing this process requires critical self-awareness, which is why he views the proletariat as a 'negative subject'. The proletariat has the potential to challenge the status quo because he embodies the object of capitalist system and the subject being reproduced by it. This dual status might enable a unique 'negative' form of class consciousness, allowing workers to recognize their objectification and thus challenge reification. Similarly, the citizen/user of cities, alienated and objectified by the new enclosed urban spaces, might be the route for building a negative subject, as long as a consciousness of this condition can be built.

Furthermore, Lukács calls 'dialectical humanism', opposing conventional humanism that assumes a fixed concept of 'the human'. Marx, according to Lukács, rejected humanism because it risks turning an arbitrary social construct into an absolute (Lukács 1971, p. 186). Instead, dialectical humanism acknowledges that humanity itself is historically produced and constantly transformed. This suggests that defining objects not only shapes social relations but also structures power and control transformations. Therefore, later in *The Changing Function of Historical Materialism*, Lukács recognized that fighting reification alone is insufficient; power dynamics play a crucial role. He cites Marx's second *Thesis on Feuerbach*: 'The question whether human thinking can pretend to objective truth is not a theoretical but a practical question' (Lukács 1971, p. 198). For this reason, in Chaps. 6 and 7, it will be important to investigate how critical practices of trying to counter these hegemonic spaces, which are not neutral, as they reify social relations, mediates interactions, and imposes inherited social conditions on future generations.

This chapter's central question, then, was: How does the design of urban spaces, through the production of things, reproduce social relations and subjects? Building on the discussed reification of social relations, the enforcement of cultural norms, the aesthetic naturalization, and the instrumentalization of space to control behaviour, we will next make a last regression on the London's privatization of public spaces, to focus the politics behind this process, and uncover how desires are embedded in things. But the question of how that is done, will be investigated in the next two chapters (Chaps. 4 and 5) by delving into the methodologies applied by designers and stakeholder (fetish) and the reproduction of subjective desires (fantasies) and its role in sustaining spatial reification.

Going Back into the Swing of Things: Politics, Desires, and Fantasy

Urban space is not merely a backdrop for social interactions but an active site of power, control, and contestation. Throughout history, public spaces have shaped and been shaped by political, economic, and social forces, reflecting broader struggles over democracy, authority, and exclusion. Contemporary urban environments, through spatial production, function as mechanisms of social control, extending neoliberal subjectivity and suppressing the inherent tensions of democratic life. Urban space might never have been fully open, but current enclosure process is forming new disciplined environments, gracefully transforming social relations and subjectivities.

Many authors argue that conflict and tension are inherent to democracy in space. Rancière (2005, pp. 18–26) stated that the *Agora* had a fluid existence sustained by the living words of its interlocutors—the breath of life. The reality of the polis itself was like a dancing chorus, like a choreography created by the changing pace of the *Agora*'s speakers. For Zukin (2010, p. 130), democracy 'in politics as well as physical space can often be loud, unruly, and unpredictable. And also dangerous'. Therefore, the calm business-friendly environment of today exists only through negation and through a soft-disciplining disguised as service to 'better serve you', which serves to hide any sign of this environment's imposed social order.

Sennett argues that the transition from Christianity to Modernity changed rituals for labour and self-discipline such that comfort was directed to home and fatigue was directed to the workplace, thereby suspending the sensory functions of the body. A passive relationship with the environment was built by the anaesthesia offered at home, on the one hand, and the disciplinary punishment of the street, on the other. Today, we observe the emergence of a new type of sensory suspension that is formed by both velvet discipline and tangled alienation.

These new public spaces extend the enclosure of the subject to the whole field of the urban experience. If, in early modernity, the social relations engraved in urban 'things' enabled the emergence of a new subjectivity in the metropolis, as well as a new revolutionary politics of the masses, the move towards an enclosed urban space extended the capitalist reification from the working space towards the total social space. It is in this sense that we argued that reification is part of the process of reproduction, always expanding its limits without changing its nature. Therein lies the reproduction of social relations by means of architecture.

In this experience, urban design operates as a veil of spectacular images that conceals the reproduction of enclosed social relations, extending capitalist reification beyond the workplace into every aspect of daily life. There, enclosed spaces are reproducing the estrangement of the subject from the means of production of their own conditions. By shaping space as an instrument of soft discipline and economic control, contemporary urban environments neutralize dissent and channel social life into regulated, consumerist frameworks. Recognizing this shift is crucial for reclaiming democratic spaces and resisting the spatial reproduction of alienation.

3.3 Desires Inside the Things

For now, we would like to summarize one last point: how desires are embedded in things. The things that surround us are not neutral; they embody desires, social relations, and historical processes. From philosophy to urban design, things function as representations of collective aspirations, shaping and being shaped by the way society organizes itself. This section explores how desire is embedded in things, drawing on Schopenhauer's concept of the *will*, and Lukács' theory of reification. By understanding the ways in which urban things materialize social relations, we can uncover the hidden mechanisms through which space is enclosed, controlled, and commodified.

Arthur Schopenhauer (1909) identified objects (or, we could say, 'things') as the representation of the 'will' (or, we could say, desire). This means that the subject frames reality to make the subject's desire possible. To identify an object in the real world is to build the means of achieving a goal (i.e. the world is formed by desires). Schopenhauer finds this conclusion in the integration of the idea (from Plato) and the a priori (from Kant). For Schopenhauer (1909, §31), the thing-in-itself is the same as 'will'. Thus, all phenomena are only representations of desire, i.e. 'objectified will'. He reaches this conclusion by asserting that the world outside Plato's Cavern is only another world of shadows (appearances). Thus, knowledge of the world resides precisely in the logic and judgements that allow the subject to predict the sequence of and relationship between the shadows, which are actually truthful to both inside and outside the cavern:

> Their wisdom would thus consist in predicting the order of the shadows learned from experience. The real archetypes, on the other hand, to which these shadows correspond, the eternal Ideas, the original forms of all things, can alone be said to have true being (*óntôs ón*), because they always are, but never become nor pass away. To them belongs no multiplicity; for each of them is according to its nature only one, for it is the archetype itself…' (Schopenhauer 1909, §31).

In this philosophical framework, we can visualize that the ideas recognizing the interaction between the shadows are actually the only reality that is always true in itself and exists throughout the different incarnations of these shadows (both the things inside and outside the cave). And these ideas are built on what Schopenhauer calls the will, the drive that mobilizes a subject to identify an object, the drive that is represented as an object. In this sense, although subjects' desires might be hidden in things, these desires are what bring things into existence. Nevertheless, updating Schopenhauer's universalist ontology into historical and materialist reasoning is fundamental. This 'will' is not a universal and abstract spirit (as reason is in Hegel); rather, it is historically produced, formed of subjects agency operating on a field of collective desires.

For instance, this shift allows us to understand how the objectification of architecture in Rem Koolhaas' Biennale is not just an isolated event but a symptom of a deeper process. In the overall scheme of things, architecture itself is turned into a reifying practice, wherein reified desires are reproduced by means of disciplined subjectivities and, ultimately, architects are objectified, as we shall investigate in the last chapters.

This investigation has forced us to confront the social reality of things. Georg Lukács (1971) developed his analysis of thingification by confronting mediated and immediate, form and content, reason and thing-in-itself, thus revealing how 'what is' is the result of a process of human production in a historical and social context. Thus, we see how urban 'things' are not simple natural objects but reflections of a broader social reality, which shapes and is shaped by the built environment. Our investigation revealed how contemporary urban things are a step forward in the transformation of social prerogatives into an objectification of the entire city, framing the experience of reality and the possibilities of its appropriation.

It is in this way that the new forms produced in London are similar to the process of primitive accumulation, in which the citizen is progressively separated from the means of production and appropriation of place, and at the same time behaviours are reified. This becomes evident in places like New Street Square, Aloft Docklands, and Renaissance, where spatial design operates not merely as aesthetic expression, but as a mode of expropriation and behavioural framing. Subjects are no longer co-producers of urban space but passive users of choreographed environments. The citizen cannot shape, claim, or meaningfully inhabit these spaces beyond the consumer roles prescribed to them—mirroring how primitive accumulation historically dispossessed people from common land and collective means of survival, while simultaneously enforcing behaviour through statutes. This process exemplifies the growing estrangement of individuals from their urban environment, leading to a form of passive participation in urban life—a new form of enclosure: an urban enclosure.

Just as the original process of primitive accumulation operated through both the confiscation of land and the criminalization of non-wage behaviours, so too does the contemporary urban condition follow a similar dual logic. Today, this logic is re-enacted through the privatization of public space (confiscation of the commons) and the scripting of consumer behaviour within them (disciplining of subjectivities and new legislation of behaviour control in public spaces). But now, the control is no longer oriented solely towards wage labour, but towards consumer compliance and acceptance of the status quo—enforcing modes of presence, movement, and interaction that serve capital through spectacle and criminalizes any dissent. This new condition of urban enclosure reasserts the role of cities in the global economy by extending the reification of social relations far beyond the realm of labour, encompassing the entire terrain of everyday experience.

This process transforms users into 'free sellers' of their citizenship through the reified mechanisms of tangled social rules, driven spatial consumption, and velvet architecture. Cosy and kind, domestic and comfortable, these are the new, emerging, bird's-nest forms of imprisonment (a new layer added to former means of control); they operate through disorientation and comfort, rather than disciplining and punishment.

What we see here is that the spaces produced are not simply the result of a neutral aesthetic or functional design, but the unconscious reproduction of broader social structures. The investigation of the production of ‘things’ in London’s public space shows that there are hidden intentions lying beneath their appearances. These things are the result—the spatial individuation—of a deeper layer in the collective unconscious. It is for this reason alone that these things seem to be objectively real. That is why Latham and Koch’s elements appear to be plausible and verisimilar; they are the result of current hegemonic social desires and unconscious. As Marx noted

> The categories of bourgeois economics consist precisely of forms of this kind. They are forms of thought which are socially valid, and therefore objective, for the relations of production belonging to this historically determined mode of social production, i.e. commodity production. The whole mystery of commodities, all the magic and necromancy that surrounds the products of labour on the basis of commodity production, vanishes therefore as soon as we come to other forms of production (Marx 1990, p. 169).

It is to this ‘mystery’ and ‘magic’ that we shall turn our investigation in the next two chapters. The products delivered by urban stakeholders are rooted in unconscious disciplined methods. Arguably, the way forward, if we are to expand Lukács’ self-critique, is to investigate the ‘means’ by which this thingification operates, in order to reveal how these seemingly neutral objects are in fact mechanisms of power and control. Although the assessment of urban ‘things’ and the deconstruction of its ‘immediate appearances’ has already provided a revealing picture, the investigation must go beyond actualized appearances because these appearances are produced by acts of power, which install a social order and perform a social function. If the shift from understanding things as passive to understanding them as active agents in the reproduction of social relations is the first step, the next one is to reveal the methods used to produce these things.

Reification becomes possible through *fetish*—the sensuous act of producing phenomena and fantasies that trick subjects into believing that inanimate objects inherently possess social performances and constructed desires. In architecture, this process manifests as the aestheticization of power, where urban forms conceal their role in social reproduction and economic control. By investigating the contradictions and struggles embedded in spatial production, we can reveal the mechanisms through which architecture reproduces these social relations, exposing its techniques of phenomena and fantasies, opening pathways for critical resistance.

Chapter 4
Fetish as it Happens: Depoliticization of Revolt and London's Housing Crisis

Abstract This chapter develops further case studies to better understand the actual process of producing objectified social relations, examining how architectural discourse depoliticizes urban production, focusing on London's housing crisis. It deconstructs urban science narratives that frame ideological viewpoints and enable expropriation practices. Using Space Syntax as a case study, the text exposes how supposedly neutral methodologies reinforce economic and spatial inequalities, legitimizing political discourse, displacement, and privatization. It also analyses how London's housing policies since 2009 illustrate this process, where bureaucratic rhetoric fetishizes private-sector solutions while masking dispossession as public interest, resulting in an unprecedented housing crisis. Based on these contradictions, this chapter regressively analyses the historical role of 'genius' figures in creative labour, showing how they shape narratives that capture collective labour. It traces the association of disciplinary forces with mystification, from priestly origins to contemporary entrepreneur stars. The chapter further interrogates the fetishization of urban design production, calling for a critical reassessment of acritical subjectivities sustaining power structures through spatial production.

Keywords Depoliticization · Political discourse · Housing crisis · Fetishization · Urban policies · Space Syntax · Expropriation · Privatization · Neoliberalism · Genius myth

4.1 Magic Blasphemies and the Depoliticization of Revolt

> It is interesting that in Ancient Egypt, and certainly in South Asia as well, the architect was privy to secret knowledge, again knowledge that would be used in constructing a temple, constructing a palace, again, building typologies really associated with the upper echelons of society and power. And so, architects were also thought, particularly in Egypt, to possess magical powers and I think in today's architecture profession they still cultivate on that as well, with the star architect. I mean, I was just reading now, he would probably dispute this, you know, identification with the star architect... the Swiss architect Peter Zumthor and the notion that he has to interview the client, and the client has to convince him that she or he is worthy of having a Peters Zumthor architecture. So, again, sort of being initiated into the

C. V. De Lima Amaral, *The Urban Reproduction of Subjectivities*, SpringerBriefs in Geography, https://doi.org/10.1007/978-3-032-01465-8_4

> cult of the priestly caste of architecture in some way. And, in a funny kind of way, I think that mixing up architects with magic and religion and, again, having certain powers still continues on today. (Interview with Mary Woods, in Arbuckle Industries 2015)

In January 2016, David Cameron (2016) identified the demolition of council houses as a key strategy to tackle social problems, to help businesses flourish, and even to improve 'national security' against 'Islamist extremism'. For him, 'the bulldozing of sink estates' was at 'the heart of turnaround Britain'. For David Cameron, the strategy was simple: to 'sweep away the planning blockages' so that the rules in play for all would not apply to the people in power; to build 'new steps' that would 'reduce political and reputational risk' so that the reputations of the people involved would not be destroyed by the outrage caused by such measures; and to create new uses for public spaces that were seen as 'wasted open space'. As a good aristocrat, Cameron saw utility only in the areas of the 'park' (for pleasure) and the 'garden' (of earthly delights), which valorize the buildings as commodities, and, of course, as he stated: 'To finance this, we'll establish a new £140 million fund', as the supposed entrepreneurs could only make this enterprise possible through the exceptional channelling of public funding (thus appropriated as private).

With that, Cameron's plan was to destroy 100 housing estates, displacing 100,000 residents. Thus, what was put into play was precisely an assault (by the rich) on the most vulnerable and poor, or, as Cameron would put it 'my second term agenda is to wage an all-out assault on poverty'; and he added: 'regeneration plans. For some, this will simply mean knocking them down' (Cameron 2016). How could David Cameron so influentially identify the root causes of all social evil in council estates? Well, he did not do that alone; he used architecture (research and theory) to support his conclusion, thus glossing it with an air of objective truth:

> The riots of 2011 didn't emerge from within terraced streets or low-rise apartment buildings. As spatial analysis of the riots has shown, the rioters came overwhelmingly from these post-war estates. Almost 3 quarters of those convicted lived within them. That's not a coincidence (Cameron 2016).

Of course, we do not aim to argue that this is coincidence. Quite the opposite: we want to determine how architecture became, on the one hand, an instrument used to manage inequalities resulting from years of segregation and lack of social investment and, on the other hand, a source of powerful rhetoric to further expropriate the poor. This is exactly what was done by the prestigious Space Syntax Laboratory at the Bartlett Architectural School, in partnership with Space Syntax Limited, a company funded to develop commercially viable applications of the work developed in the Laboratory, 'to provide industry-focused research for commercially-minded clients' (see www.spacesyntax.com).

Space Syntax aims to use the technology of mapping and data processing in urban spaces to improve the 'operational performance' of these spaces. Thus, the company uses a series of techniques to conduct simulations and the spatial modelling of data, to forecast models and predict human behaviour, to produce highly graphic visual analyses and find patterns and geometries, all 'evaluated using a mathematical algorithm' and with 'the use of our evidence-informed tools and techniques' (see 'service offer' in www.spacesyntax.com). In this way, its research decisions seem to be the products of an automaton, with no ethical choices involved; they seem to be based only on 'objective truth'. Arguably, this rhetoric is not a straightforward lie, it is a form of 'fetish', creating concrete reality in a realm of simulacra.

In its report titled *London, Riots Research*, the Space Syntax (2011) group eliminates all variables that might interfere with the model it has built—for instance, access to good quality health and educational services—so that the result of the research coincides with the pre-existing hypothesis, namely 'a hypothesis that riot incidents occurred in places that were both near town centres and near large post-war housing estates'. As we saw in Lukács' and Marx's discussions about the fetish of things, once you take the immediate as an absolute, thus eliminating any critical considerations of the whole of society (say exclusion, education, inequalities, racism, and so on), the reification of reality becomes effective truth.

In addition to the astonishing complexity of the 'algorithms' deployed, which makes viewers feel they are witnessing a magic show, the studies of the Laboratory generally deploy a simple method based on the theory that 'Space is the machine' (Hillier 2007). First, a total reification of space into a trivial machinist system is assumed; then, convoluted methods and vast amounts of collected data are presented; then, some type of automaton algorithm performs calculations that are impossible to understand (always summarily mentioned, as they are too complex for a detailed explanation, and sometimes presented as a short video of moving objects); then, the rendering of models, maps, and astonishing graphics makes the simulation seem more real than reality itself. After this careful manipulation of information, the conclusions (especially the raw and simplistic ones) magically confirm the first 'hypothesis'. Thus, any conclusion is presented as the revelation of an invisible truth.

It is in this sense that the production of truth cannot be separated from its ethical and political implications because these implications are the sources of that production, even when they are presented as impartial. Douglas Spencer (2012) demonstrated how hegemonic architectural discourses on complexity are engaged not only in legitimating but also in constituting the special means of post-disciplinary societies of control (2012, pp. 3–11, 98). For instance, in his account of the BMW headquarters designed by the Zaha Hadid Office and the Ravensbourne College designed by FOA, Spencer systematically unveils how contradiction is wiped off the concept of complexity. In the case of the BMW project, this is done by an attempt at 'managing' and 'articulating' complexity by naturalizing humans as 'molecular' agents subjected to self-organization in a 'laissez faire naturalism', which is carefully produced by architecture to apply the general means of a swarm theory, thus producing a collective subjectivity passive enough to be captured (Spencer 2012, pp. 41–67).

In the case of FOA, the office conceives 'complexity' as a sort of universal Hegelian ideal emerging as a natural process without agency (Spencer 2012, pp. 131, 134). In so doing, the office conceives a theory of complexity that eliminates politics, history, criticality, and any conceptual speculation. In this way, complexity can become a means of steering and control in the foundation of a sort of 'ordoliberalism' (Spencer 2012, pp. 122–163). In FOA's project for the Ravensbourne College, the conditions of contemporary capitalism (such as precariousness, market realism, citizens as consumers, fragmentation, deterritorialization, and networking) are designed into the building spaces, aiming to naturally frame the students' subjectivities.

In this sense, fetish does not simply 'hide' truth as if it were lying. Fetish actually creates reality in a very specific way. It not only provides a logic for the functioning of objective reality, but it does so by means of mysterious artifices operating through the created objects. In this sense, architecture not only hides intentions within space, but it also reproduces them, as the viewpoint—framed into the reality—further frames the subjectivity of others, as if it were a magical power revealing the natural character of existing reality (see Mary Woods quote above).

In this chapter, we aim further to analyse the use of fetishist rhetoric through concrete examples of housing policies in London; so we can perform a regressive analysis of the operational use of fetish in discourse about creative labour, especially labour producing new urban forms, highlighting how its roots are associated with the disciplinary forces—an orchestrated carnival of pirates, courtiers, and stewards (to make a reference to a Foucault expression we will come back on the next chapter). This picture aims to trace down strategies and methods of how fetish not only operates in the production of 'external' objects but also producing an objectification of the subjects doing the creative labour. This will allow to formulate, in the next chapter, a definition of fetish and discuss how phenomena techniques and fantasies builds on contradictions of unconscious desires. The point is not to blame architects, entrepreneurs, or urban stakeholders for how things are; rather, we aim to reveal how these reified subjectivities come to inhabit their unconscious. Or, as Adorno explains this goal,

> Dialectics is the self-consciousness of the objective context of delusion; it does not mean to have escaped from that context. Its objective goal is to break out of the context from within (Adorno 2004, p. 406).

4.2 Deconstructing Fetish in the London's Housing Crisis

London's housing crisis is not merely the result of economic imbalance or demographic pressure—it is the spatial expression of deeper political strategies and ideological commitments. Since 2010, a succession of policy documents has framed housing as a technical issue to be solved through market-led mechanisms, masking the role of state intervention in facilitating accumulation and dispossession. The formulation of the housing problem in specific terms resulted in the imposition of a large-scale-capital solution, usually supressing the interests of common citizens. This section revisits the foundational policy documents that set this trajectory in motion, exposing how language, omissions, and strategic edits constructed the narrative to imposing an heteronomous control. By retracing the production of these texts, we reveal the devices of how London's housing policy became a theatre of fetishized solutions systematically undermining the right to the city.

To assess this process of fetishizing the housing problem, this study turned to the roots of the 'housing' policies proposed for London. Arguably, the intensification of London's housing crisis could be traced back to a 2010 policy framework (Johnson 2010), followed by a 2013 consultancy draft and a final 2014 statutory version. The 2014 version is an especially tricky and artful text, wherein statements at the beginning support contradictory proposals in the middle. Furthermore, a careful analysis comparing the different versions of the documents (the consultancy document and the approved document were compared using the tools of word software) exposes a series of key changes. Since then, even if the housing policies have changed, the crisis has not—suggesting a structural continuity in how the problem is framed.

Although none of the proposals underwent any substantial changes, and many of them were amended on an ongoing basis, two facts are especially important. On the one hand, part 5, 'About this Strategy', which revealed the subjectivity of choices made in the proposals, simply disappeared, as if the proposals were the result of mysterious technicalities. On the other hand, a vast series of sentences and paragraphs—direct and revealing statements regarding the chosen strategies—were altered into obscure, labyrinthine rhetoric. The following analysis of this procedure will reveal the devices used by this politics of appearances and disappearances—a politics that governs the production of urban space.

Here, we encounter what we conceived as a fetish by *sprezzatura* (a concealed craftsmanship, a hidden skill behind the appearance). There are many examples, but due to the constraints of this chapter, only one will be presented. For instance, the sentence 'The GLA is particularly keen to explore the concept of a London Housing Bank focused on large-scale developments to generate additional supply' (Johnson 2013, p. 39) was changed to:

> The Mayor wishes to develop a London Housing Bank with the aim of accelerating the pace of development and generating additional housing supply, particularly on large sites where speed of delivery is significantly constrained by the traditional house-builder business model (Johnson 2014, p. 68).

When artifices are consciously managed, arguably, we enter this realm of '*sprezzatura*', an essentially modern technique of concealing artifice. In the opening chapter of Manfredo Tafuri's (2006) magna opera on the Renaissance, he discusses the concept of *sprezzatura*. Mentioned in Castiglioni's book *The Courtier* (about the etiquette of the court), *sprezzatura* is the main aesthetic attitude expected from a courtier or artist. This procedure consists of creating an 'illusion', just as the 'perspective' creates one in a painting, but it is applied to various realms. For instance, a courtier's 'gait' was of the utmost importance for social distinction; this gait entailed the transformation of regular walking by a series of small variations to create grace, lightness, structure, delicacy, and grandeur. However, all transgression of a normal walk should be concealed, so it would not look fake. Thus, a product could keep the appearance of being 'naturally artificial'. In the final result, an elegant courtier gait 'looks effortless' and distinct. Fetish and *sprezzatura* have in common this ability to produce phenomena by hiding artifice. In the social realm, revealing the means of this type of production demands a form of social struggle.

A further analysis of the documents reveals many such *sprezzaturas*. Of great importance is the account of the Localism Act of (UK Government 2011) that transferred planning powers to the Great London Authority (GLA), 'while removing the legal restriction on the mayor to spend money on housing' (Johnson 2013). In this case, the policy builds the discourse on public investment in the private sector to legitimize the creation of the 'London Housing Bank' as a funding body of the private sector. This is one of the main strategies in London's housing policy: 'due to the imperative for increased supply of homes of all tenures, it is expected that cross-subsidy will primarily come through the provision of open market housing' (Johnson 2014, p. 70). 'The mayor wants to accelerate delivery, bringing new buyers into the market via a Build to Rent fund, backed by the government's debt guarantee' (Johnson 2013, p. 48).

In a context of financial crises, the risk of such large investments is not in the banks' interest. Therefore, it is proposed that all the risk should be assumed by the public sector, while the profit remains in the hands of the developers. This is the case because the loans will be 'available at below market rents for a fixed period, at the end of which the homes can be sold on. The funding provided by the public sector would be repaid, potentially including a value uplift' (Johnson 2013, p. 39). Therefore, developers are granted the use of public money as capital and will only 'potentially' pay it back with some uplift; but they will surely profit with zero risk:

> [T]he London Housing Bank should initially utilize up to £200 million of this funding (…) There will be flexible loan terms at lower than commercial rates, to enable registered providers to offer sub-market rents for the period of the loan. At the end of the loan period the Mayor will expect a positive return on the loan and the initial capital to be paid back (Johnson 2014, p. 69).

Furthermore, the Localism Act 2011 created a mass transfer of land ownership to the GLA, which became the largest public landowner. Therefore, the policy document has a specific section on how to privatize this public land, entitled: '4.12 Bringing forward public sector land'. It aims to do so 'by making it faster, easier and cheaper for public landowners to bring forward land for development. It establishes a framework agreement of 25 developers and contractors' (Johnson 2014, p. 77). In addition, the 'housing zones' would create 'lighter touch planning and effective land assembly, including, where necessary, the use of compulsory purchase powers' (Johnson 2014, p. 77). In her analysis of this trend, Raquel Rolnik notes that this easy access to planning permits without any control of land speculation is leading to further land banking (UN 2014, p. 10).

This new 'lighter touch planning' (so dear to neoliberal rhetoric) is (not so) surprisingly accompanied by an intensive mobilization of the state apparatus to benefit big industry: 'The Mayor will use his full range of powers to get more homes built in all sectors, and this will involve new initiatives to secure additional finance, bring forward the land for development and build the industry capacity' (Johnson 2014, p. 24). If assessed fragmentally, this statement could seem to indicate that the government is somehow retaking public control over the development of the housing sector. Nevertheless, the majority of new houses will be produced by the private sector as supposedly 'affordable houses'. Understanding the 'unaffordability' of these 'affordable' houses should be sufficient to unveil this *sprezzatura*.

In the Barnet Council case, original two-bedroom flats were offered at a value of £175,000 to those from whom they were being compulsorily purchased, while the new one-bedroom 'affordable' flats were expected to be sold for £415,000. In addition, the proposed scheme would provide 199 fewer 'affordable' houses compared to the number of social houses previously available (Booth 2015). Furthermore, former council tenants were not given any choices other than buying a new flat or facing the private rental market. This made it impossible for the local community to participate in the 'sustainability' of the development. This situation is similar to the city's other regeneration schemes. In the case of the Elephant and Castle regeneration scheme, 'affordable rents' are expected to be double the price of previous rents, and new flats cost almost three times what the originals cost (Wainwright 2014).

In reaction, local residents created an association called Better Elephant to fight this process of dispossession using live information about the social effects of the new policies. The diagram below (Fig. 4.1) was produced on the basis of a map created by the association to illustrate where the scheme was displacing current residents (Better Elephant 2013).

In their efforts to provide land for development, the creators of these strategies consistently articulate tactics of compulsory purchase and land assembly that overlap the local scale. All of the solutions produce space on a scale that overrides the local level. As Zukin (2010, p. 222) asserts, we may have not seen the end of history, but we are certainly seeing the end of 'place-bound culture'. Therefore, there is a different housing problem that does not fit the rhetoric of that policy document.

The problem that needs to be solved is not that of producing as many expensive flats as possible; it is that of providing housing for Londoners. The reality that is not addressed by any of the proposals is that of the crude living conditions in London. Despite efforts to rectify these precarious conditions, data from Shelter (2015b) show that 20.6% of the houses in England are 'non-decent' (i.e. houses in bad physical condition with 'modern' facilities that are failing, not thermally adequate or fall below minimum standards). Furthermore, 35% of private rented households in England fail to meet the Decent Homes Standard (UN 2014, p. 8). In addition, according to ONS, 11.3% of London households are overcrowded (GLA 2014).

To 'bring forward land' for development, the proposed tactic was to change the laws governing 'compulsory purchase', 'land assembly', 'anti-social behaviour', and 'affordable housing'—a concept that imposes a substitute ethics on and hides a new logic in place of council housing strategies. The result of these changes was the dismantling of traditional communities and urban places through a process of accumulation by dispossession—a term used by David Harvey (2004), to acknowledge how 'new' imperialism uses laws to strip local citizens of older forms of wealth to which they held title—and a subsequent re-concentration of property in the hands of big real estate actors. In other words, in a time of financial crisis, architecture became a tool for transforming individually dispersed patrimonial capital into

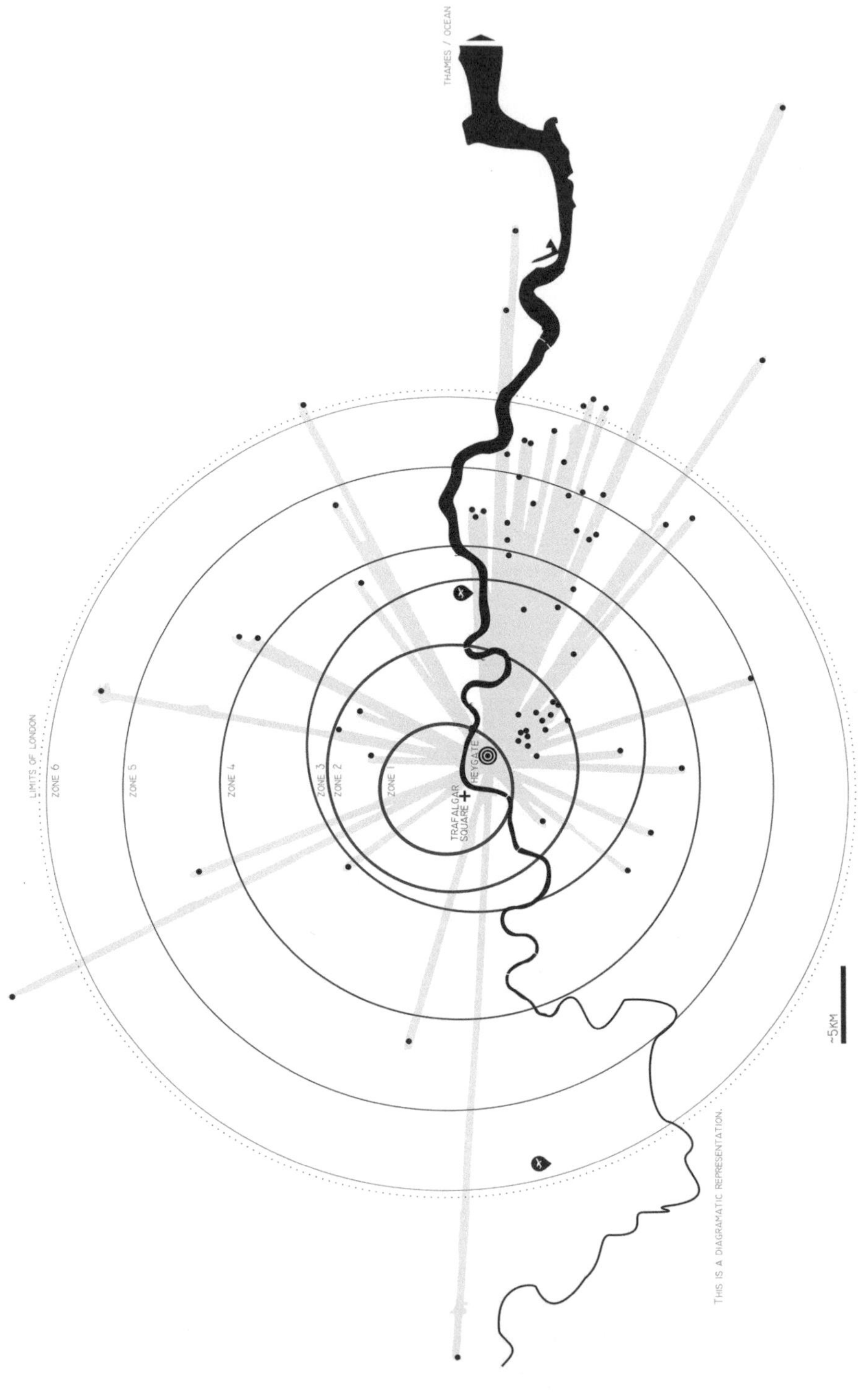

Fig. 4.1 Displacement of Heygate Estate Residents. (Source: the author (based on information from Better Elephant 2013))

concentrated capital for a centralized renting economy. Historical data collected by the organization Shelter (2015a) demonstrates that the 1980s witnessed a significant decrease in the construction of social housing in England. The annual construction rate of social housing is now 90% lower than the rate of production in the mid-1950s.

In this context, the use of Compulsory Purchase Orders (CPOs) has been advocated as the foremost tool to be used in producing a greater number of houses. An analysis of that will reveal the fetish of this proposal. Although CPOs are a legal instrument that has existed since 1961 with the purpose of enabling authorities to serve public interests (Department for Communities and Local Government 2005), the large-scale use of CPOs as an instrument of housing production in London is a trend. Evidence of this change is seen in the creation (in 2002) of the Compulsory Purchase Association, a member organization that aims to guide and promote the practice as well as to support relevant expertise and networking. The result is that the majority of local authorities make use of 1 or 2 orders per year, for a total of more than 40 per year (Dees 2012). This instrument was endorsed by the former Mayor of London (Johnson 2013, p. 77) as a key strategy for 'land assembly'.

Furthermore, the major shift in the use of this instrument is related to a shift in interpretation. Since the economic crisis of 2009, any benefit to the economy has been interpreted as 'of public interest'. In this sense, an instrument that was intended for the common good has been oriented to favour the market. In the case of a regeneration process in Barnet Council, which faced massive street protests and resistance from local residents, the council leader, Richard Cornelius, reportedly said,

> The regeneration project will transform the estate and will provide high quality and attractive new homes at a time when public finances are incredibly tight. We believe that this is very much in the public interest (Booth 2015).

Another common justification for this process is the austerity imposed upon local governments. The regeneration process is often connected with the process of privatizing public spaces, a process that saves local authorities from the maintenance of streets and squares. When asked whether the new policies on creating housing using CPOs could be implemented in public spaces in a manner consistent with the public interest, a Great London Authority higher officer—who asked to remain anonymous—answered as follows:

> It could be coherent with it. But it will not. It is not the interest of housing developers. It is not just a problem of design. The Planning Department will not reinforce it. (...) The public management of the streets involves funding. So it needs money. (...) There is a large amount of things to be done in the public space, and it is simply easier to do it with corporations and a local authority. When there is many landlords involved, things becomes difficult, and it is hard to negotiate all the interests. With a local authority holding the land, it is easier. There is just no money to do otherwise (Anonymous 2014, in an interview to the author).

In addition to this point, data from Shelter (2015a) demonstrated that from 2003 to 2014, the number of rented social houses remained nearly stable. After 2003, there was an increasing shift from houses being rented by local authorities to houses being rented by housing associations. These associations are non-profit organizations; thus, they represented one more step in the downturn of the welfare state. Furthermore, the National Affordable Housing Programme, created in 2008, is a step away from the previous models; in this new programme, public funding is allocated to partner developers to deliver homes for both rent and ownership at 80% of the market price. This policy ostensibly provides socially committed houses. Nevertheless, with a 72% increase in the ratio of the median house price to the median salary between 2008 and 2014, this 20% discount offers little to people excluded from the market. That reveals how this rhetoric is nothing more than an instrument of wealth dispossession and property accumulation.

Thus, London's housing problem is until now grounded in this illusionistic basis, and the proposed way of managing the crisis creates a 'false solution'. The assumption of the 2014 London's housing strategy was that an increase in private production was the solution that will enable the delivery of necessary housing. Although 77% of the production of new homes was already in the hands of the private sector, overall production remained at less than half of the baseline needed (UN 2014, p. 7). Therefore, London's policies were actually disenfranchising the city's most vulnerable residents of the right to adequate housing (UN 2014, p. 20) while producing a massive process of property accumulation by dispossession. This reveals a fetish that empowered a 'naturally artificial' process of dispossession.

In 2022, the Right to Buy extension to housing association tenants was widely criticized as a 'gimmick' that would worsen the access to housing (Wilmore 2022). More recently, Sadiq Khan's 2025 green belt land release plan seeks to add homes but raises concerns over erosion of environmental areas and reinforces market led solutions for social housing (Adu and Horton 2025). The continuation of the affordable housing scheme remains a rhetoric that keeps the problem within the market realm, in a context where most of the population does not have access to million-pound houses. Other critics note that: 'There is a vital point at the core of this issue. Even if Starmer has often given the impression that the answer to the housing crisis lies in clearing away planning law and letting corporate developers do the work, their ring-road faux-Georgian cul-de-sacs will not provide anything like the entirety of the solution' (Harris 2025). These policies exemplify the contradictory dynamics of the urban housing crises: on the one hand, the plans challenges traditional land preservation policies to address affordability; on the other hand, it risks reinforcing neoliberal trends by reframing green spaces as exploitable assets and housing rights as commodities.

Furthermore, what emerges is not simply a distortion of housing policy, but a broader ideological mindset embedded in spatial reasoning. These developments underscore that London's housing policy is not just contradictory, it is actively accelerating neoliberal spatial restructuring, increasingly converging with mechanisms of protest control. That redefines not only who has access to housing but who can occupy space politically. The erosion of democratic space is no longer metaphorical: it is measurable in protest bans, surveillance regimes, and spatial exclusions. These tendencies, show how the subjectivity of architecture and urban planning is governed by both market rationalities and a deeper aspiration for power, public control, and silencing of differences.

In synthesis, although David Cameron's use of Space Syntax Laboratory simulations was a means of imposing a truth, and Boris Johnson's *sprezzatura* was used to create a false problem, and Keir Starmer's zoning pragmatism is serving to present itself as technocratic neutrality (while maintaining the previous status quo of market-based solution and environmental instrumentalization), those are only three steps in the long history of using architecture as an instrument of fetish in social space. How deeply into the discipline of architecture do these fetishist procedures go? Are these magical phenomena of architecture isolated cases, or do they have roots in a broader social experience?

We can envisage a conflict of logics emerging from these black boxes of neutral theories and methodologies. There is a conflict between fetish as a means of reproducing social relations and the events that (ethically) create such relations as the only possible alternatives. In addition, we saw in previous chapters how this urban condition surpasses gentrification (the exchange of one class by other) and entails the reification of the whole experience of citizenship in the city. Next, we diverge into a regressive analysis of disciplinary forces framing the subjectivity of creative labour with the aim of revealing the fissures in its linear narrative of neutrality. With that, we aim to further reveal the roots of fetish, and the fundaments of the theatre that make it possible in the first place.

4.3 Regressions in the Production of Fetish

As we saw above, the production of urban 'things' presupposes an image of the whole or, more precisely, institutionalizes a way of conceiving truth and society. Thus, the production of things is simultaneously the reproduction of specific forms of social relations. The architect functions as a gear in the manufacturing of these social orders. As we go deeper into our analysis of fetish in the discipline of architecture, we shall recognize that fetish not only reproduces social relations and produces things, but it also reproduces the discipline of architecture itself and produces an architectural reified consciousness. Thus, the production of truth in architecture, in addition to having an implicit conception of society, also has an implicit conception of what architects are and how architectural production happens. Next, we aim to investigate regressively the mechanisms involved in picturing the 'geniality' of the architect, who supposedly acts as a super-individual—conceiving and creating social space.

Mary Woods (1999) termed this vision of a super-individual architect as 'Roarkism', after the character Howard Roark in Ayn Rand's novel *The Fountainhead*, which 'focus[es] on the architect as solitary creator to the exclusion of other narrators and narratives' (Woods 1999, p. 1).[1] Thus, Woods aims to bring the *mise-en-scène* and the background to the forefront in order to understand the formation of

[1] Adam Curtis (2002), in the documentary *The Century of the Self*, captures how—progressively over the last century—the narrative of the self, and the individual's self-realization, spread as a dominant ideology in connection with a change in the mode of production and a new focus on consumerism and the realization of personal desires.

the profession of architecture; she investigates the conflicts and the multiple participants that support the actions of the architect (such as the schools, the building committee, the courtroom, the clients, the admirers, and the critics). In doing so, she aims to break with the traditional narrative of architectural history—a narrative based on the architects' biographies (as we discussed in Chap. 3) that has little to say about how architecture is actually done. However, despite Woods' intention, over the course of the book, the individual figures of the first professional architects in America continue to play the role of guiding lights in her historical fabric.

The above observation is not, in itself, a depreciative comment on her work; as we have seen, fetishes have a real impact on reality. Nevertheless, next, we aim to investigate how, besides using fetish to forge an implicit vision of the whole, the discipline also produces a specific form of fetish: the 'architect' themselves. Once we reveal the fetish creating the image of the architect, we shall uncover the underlying fantasy (the third and supporting layer in our unconscious architectural pyramid).

This is not, however, an eccentricity of the figure of the architect (although it is not a coincidence that it was the chosen subject of Rand's 'bible' of individualism). As in all reifications, the whole of society is constitutive of its parts. Thus, we shall use examples of major figures of fetish in the contemporary mythology of 'genius' to develop a critical awareness of those types of social 'facts'. We shall start by recalling the most emblematic figure of the 'genius' in our times, namely, Steve Jobs.

Biographies of Steve Jobs (*Jobs* 2013) usually present the narrative of a brilliant mind who leads a revolution, and the narrative continues: despite facing much difficulty, indifference and resistance, in a magical moment induced by an LSD hallucination, Jobs had a vision of the future and worked hard to realize it. Nevertheless, the actual operative means of production he created functioned through propaganda. By constructing a magical narrative, he was able to recruit brilliant minds to work in his name. Furthermore, the actual elements of innovation (the GPS, the voice-activated system, the touch-screen display and the internet) in his most famous 'invention'—the iPhone—were actually developed by state-funded research (see Mazzucato 2011). Thus, the fetish of Jobs was not to mysteriously create the iPhone; rather, it was the creation of a company/narrative able to appropriate both the work of individuals and the social work developed by state-funded research. Thus, the secret of his success was the creation of this accepted mythology that he was himself a genius. By this means—the fetish of the genius—he concretely managed to appropriate the work of others as the result of his own genius.

Furthermore, Jobs did not invent this abstract machinery for expropriating inventions. Thomas Edison is perhaps the most iconic modern 'inventor' and has been systematically depicted in schools and by the media as the genius behind the invention of such things as the 'light bulb'. Nevertheless, a series of studies of the actual means by which the inventions (commonly attributed to Edison) were produced shows a different picture. In the early days of the Thomas Edison laboratory in West Orange, he was employing more than 200 scientists, craftsmen, labourers, and machinists. When the laboratory expanded and became associated with a factory complex, the number of employees jumped to 5000, and today General Electric employs more than 300,000 people (Padgett 2016). At the beginning, these men were paid only 'working man's wage'; however, the famous inventor reportedly said that—in exchange for their ambition—his employees were given the opportunity to work side by side with a genius (Bellis 2016). However, how much of the lab's creative labour came from Edison and how much came from his workers?

Carlson (1988) has studied the process by which the alkaline storage battery was invented. He argues that during the course of this invention, Thomas Edison developed a new way to produce inventions. Previously, Edison would work with mechanics and craftsmen in a relatively loose way, wherein those workers would investigate diverse aspects of an invention, and eventually Edison would step in 'only at the appropriate moment' to 'pull together the various discoveries and improvements into a successful invention'. Later, in Edison's laboratory, large groups of chemists, engineers, and college-educated scientists would work on experiments focused on very specific goals, in a systematic, step-by-step arrangement of assignments. In shifting from a 'divergent' to a 'convergent' style, Edison became a manager who oversaw the project, a role that left him time to focus on strategies of productivity and commercialization.[2] However, the 'most important' role now played by Edison was to 'motivate his research team', using 'decidedly informal' techniques of 'motivating and directing' through his 'use of a personal, folksy style [that] may well have been deliberate' (Carlson 1988, pp. 10–11).

> Although this convergent approach produced highly reliable results, it came at the cost of requiring over 50,000 individual experiments. Furthermore, Edison had thoroughly routinized the innovation process. By breaking down the research into a sequence of small, standardized experiments, Edison had altered the creative process from hands-on ingenuity and skilled observation to persistence and careful record-keeping. Gone were the last vestiges of the 'heroic' myth of invention in which insight came in a blinding flash; results now came by plodding through innumerable experiments (Carlson 1988, p. 6).

Invention became an 'orderly', 'predictable' process, making Edison's 'large staff and substantial facilities' an advantage to beat competitors, which in turn made 'the innovation process a reliable component of business strategy' (Carlson 1988, p. 11). For these reasons, his friend Henry Ford reportedly said, 'Mr. Edison gave America just what was needed at that moment in history. They say that when people think of me, they think of my assembly line. Mr. Edison, you built an assembly line which brought together the genius of invention, science, and industry' (quoted in Newton 1987, p. 31).

For these reasons, the same critique Marx applied to the fetish of the commodity (used by capitalists to alienate products from workers in industrial assembly lines) can be applied to the fetish of the invention (used by the 'genius' to alienate the creative work of a collective). Correspondingly, if Thomas Edison was famously fond of saying 'Genius is 1% inspiration and 99% perspiration', we could say that what was actually taking place was 1% fetish and 99% expropriated work.

In addition to developing this abstract assembly line to expropriate the work of others, Thomas Edison cultivated his fame through vast investments in marketing

[2] For instance, he participated in the association that developed 'programmed obsolescence'.

and especially through the legal mechanism of patents. Edison alone is credited with the invention of 1093 patents (Simonton et al. 2015). Lemley (2011) investigated how the conception of a sole inventor—which is implied in patent law—is a myth, as inventions come from progressive collective work and are therefore frequently produced simultaneously by independent groups, as 'Inventors build on the work of those who came before, and new ideas are often "in the air"'. Thus, a patent is also a means of privatizing the work of others.

That said, we should not be at all surprised by the fact that the most mythological figure of modern science was working in a patent office in the year he published a series of 'groundbreaking' papers, causing scholars to nickname that year *annus mirabilis* (see Rynasiewicz and Renn 2006; Sallent Del Colombo 2013). Indeed, Albert Einstein was well versed in the art of the patent (the assembly of small step-by-step innovations into a big invention). When he finished his university studies with low grades, he did not find a job as a researcher and was left working as a clerk in a patent office, with no research group and no assistants (Contieri 2005).

Nevertheless, Einstein's papers—quite unusually—contained no references at all to previous studies, a finding that has recently raised a series of doubts regarding credit for the theory of relativity (Gine 2010). Although Einstein did not reference Poincaré with regard to the conception of relativity (Darrigol 2004), he also failed to mention the source of his famous equation—whether it was from the Englishman Oliver Heaviside, the Frenchman Henri Poincaré, the Italian Olinto de Pretto, or perhaps all of them (see Bartocci 1999; Drageset 2015; Socolovsky 2012). The point here is not to suggest that he was simply a 'plagiarist', as many have recently argued. Rather, he was just a 'scientist' doing the regular work of a 'genius' by presenting the result of a socially constructed theoretical development as his own magical invention.

Perhaps the most remarkable aspect of this event is the fact that while Poincaré was capturing the work of a legion of disciples in his academic work and managing his distinct position in the academic field, Einstein was able to surpass Poincaré through a new modern form of fetish enabled by the political context of struggle between empires. In this context, Germany was struggling to emerge as a dominant power as tensions were building between the end of the nineteenth century and the breakout of World War 1. Thus, although Einstein had no position in the academy—which means no support from academic corporations—he was supposedly able, in a short period of time, to publish a series of papers 'revolutionizing' physics. Furthermore, the 'mysterious' (Darrigol 2004) similarities between his theory and Poincaré's were simply ignored not only by the German journal *Annalen der Physik,* which published it, but also by the enormous machinery of propaganda erected at the dawn of media and cinema in the years immediately before World War 2. At this time, Einstein was exiled to the emergent power of the United States. In this way, Einstein emerged as the first celebrity scientist, as 'famous as Chaplin', and he became symbol of a pure (detached and objective) science in opposition to the ideological Nazis.

Again, this is not to say that relativity is false, nor that Einstein was lying about relativity. Perhaps he was not lying that he had formulated the theory, as he likely believed in his own fetish. However, precisely because of that belief, it is possible to see how the fame of this iconic figure was produced upon a fetishist device (the patent) and to see how his own status as a genius was produced by another fetishist device (the media)—two devices that precisely composed the main geopolitical power of the emergent empire.

Again, it must come as no surprise that one of the chief theoreticians of the United States' empire was also an architect. Thomas Jefferson once asserted: 'I am persuaded no constitution was ever before so well calculated as ours for extensive empire and self-government' (quoted in Negri and Hardt 2000, p. 160).

Here, Jefferson was praising the construction of what he termed the 'Empire of Freedom', where power was to be exercised over others while internal democracy would construct an apparently natural state of justice. Thus, freedom in this empire has distinct meanings—whether you consider it to be used inside or outside—while equality and fraternity were not even part of the formula. Tocqueville interpreted the foundation of this new form of power as 'soft despotism', in which the actual means of power were exercised using deceitful—as opposed to frank and open—methods of violence.

Noam Chomsky (2003) analyses the formation of this empire to understand both the contemporary state of affairs and what would ultimately lead it to become a Terrorist State, promoting wars against non-aligned states in order to impose rules, laws, and means of production (in short, the means of fetish). As a metaphor, he asserts that behind the mask of the Emperor, there is a Pirate:

> St. Augustine tells the story of a pirate captured by Alexander the Great, who asked him 'how he dares molest the sea.' 'How dare you molest the whole world?' the pirate replied: 'Because I do it with a little ship only, I am called a thief; you, doing it with a great navy, are called an Emperor' (Chomsky 2003, p. vii).

Thus, the fetish of empire is only a more sophisticated means of expropriation. Nevertheless, this is not a completely new invention. The British Empire already ruled a vast area 'through and by the natives' (Frederick Lugard) using the symbolic power of coercion to impose its rules and laws on colonies, as in the famous couplet 'Whatever happens, we have got the Maxim Gun, and they have not'—the Maxim Gun was the weapon most associated with the British imperial conquest, and it was patented by the former American and later British inventor Hiram Stevens. Hiram Stevens was also famous for litigating in court against Thomas Edison over the patent of the light bulb. These episodes, in which the truth of an invention is litigated, are not exceptions to the context of wars. There are many such cases: the litigation between Newton and Leibnitz over the invention of the differential calculus, or Santos Dumont and the Wright brothers over the invention of the airplane, or even the centrality of Darwin in the 'invention' of evolution. All of those can only be explained by the analysis of Empires fighting for positions of power, which allows imposing their fetish.

Investigating architecture in Thomas Jefferson's Empire, Mary Woods (1999) unveiled the formation of the 'professional architect' in America, replacing the former version of gentlemen architects such as Jefferson himself. Instead of coming from a noble background, this new kind of architect emerged as a modern professional, situated in between clients and builders. Nevertheless, Woods asserts that Jefferson personally supported many such architects, and the profession was gradually constructed outside the realms of the old elite, thus developing into a new field of liberal professionals. A new (liberal) Empire creates a new (liberal) form of fetish.

The above discussion comes to a high point that closes a circle. As we mentioned above, Foucault asserted that it is not the case that knowledge is power; rather, power is the position of defining knowledge and truth. Thus, those in positions of power (be they nobles in the courts, Boris Johnson or David Cameron) are able to impose a specific *sprezzatura* of reality. This circle brings us back to the first quotation of Mary Woods. In Egypt, architects were part of the royal court, and the very first figure of an architect was Senenmut. Senenmut was a former Steward of the Queen, a position that developed his knowledge and allowed him to serve members of royalty by transforming fetish into concrete architectural objects, the fundaments of Egyptian civilization. Thus, the re-emergence of the new class of butlers in London (Batty 2016)—highly educated professionals occupied with serving the 1000 'family offices' and managing the wealth of the new superrich from inside their houses—should come as no surprise. The circle is closed: the new geniuses are themselves objects, servants of the status quo.

Despite the advances of her research, in an interview, Mariana Mazzucato declared that she was not saying that Steve Jobs was not a genius but rather that one should recognize the bigger picture of how innovation is really developed in society. With this analysis, we show that one must go a step further and assert the unpleasant consequences of these findings. Steve Jobs was not a genius who invented the iPhone, he was a representation of a collective desire, a narrative able to capture the work of others as his own. This representation is the figure of a Prince in Althusser's (2000) sense and to be further developed in the next chapter closing with the idea of fantasies. However, this point should be sufficient for now to close the topic of fetish in architecture.

Chapter 5
For a Theory of Urban Fetish: Techniques of Phenomena and Fantasies

Abstract Building on the dialectical methodology used in previous chapters, this chapter develops the case study topics within a theoretical framework. It draws on scientific epistemological critiques and Neo-Marxist perspectives to examine how facts are socially produced and accepted through mechanisms of fetish, referred to here as techniques of phenomena and fantasies. The chapter explores how scientific controversies and social distinctions have been objectified in "black boxes" of methods and techniques, thereby obscuring their underlying assumptions. As a result, even practices that aim to critically challenge the *status quo* end up reproducing specific subjective senses. This analysis highlights the ways in which scientific practices, despite their intention to question norms, can become entangled in the very structures they seek to critique, ultimately reinforcing the social order rather than disrupting it.

Keywords Scientific epistemology · Social production of facts · Fetish · Factishes · Techniques of phenomena · Fantasies · Black boxes · Autonomism · Abstract machines · Production of subjectivities

5.1 The Fetish of Facts

> The good historian, the genealogist, will know what one must think of this masquerade. He must absolutely not refuse the spirit of seriousness, but instead he should want to take it to the extreme: he wants to stage a big carnival for the time when the masks return incessantly. [...] genealogy is a history just like an orchestrated carnival (Foucault 2005, p. 278).

In the quotation above, Foucault was emphasizing the mechanism through which he investigated the history of ideas as a succession of truths produced socially and emphasized as reality by means of power. For him, the question was not whether knowledge is power; rather, power was the position from which subjects could impose a notion as truth. In this sense, the concepts accepted by society as truths have a history of creation and struggle, which occurs through the institutions of society.

C. V. De Lima Amaral, *The Urban Reproduction of Subjectivities*, SpringerBriefs in Geography, https://doi.org/10.1007/978-3-032-01465-8_5

When Bruno Latour went to a laboratory of Neuroendocrinology (Latour and Woolgar 1986) in 1979, he entered the world of scientific truth production as an anthropologist who was entering a mysterious primitive tribe, trying to understand its activities and how these activities would produce the tribe's beliefs. He would soon unveil a world full of veiled hierarchies, machinations, and procedures for the creation of new phenomena. He noted:

> It is thus possible that there is some useful similarity between Marx's [...] notion of fetishism and the notion of scientific facts. (Both fact and fetish share a common etymological origin.) In both cases, a complex variety of processes come into play whereby participants forget that what is 'out there' is the product of their own 'alienated' work (Latour and Woolgar 1986, p. 259).

Later, Latour (1999a) would develop the idea of 'factishes'. While a fact is supposed to be a property of an independent object, and a fetish is supposed to be the projections of a subject's idea upon the things-in-themselves, the idea of factishes presupposes an indivisibility between object and subject, with the product being the result of an action, an event, where the subject acts upon reality and reality acts upon the subject. However, it is exactly because factishes are produced that they have real effects on humans, on things, and even on nature.

Thus, in Latour's sense, fetish is a fact of human life. Again, fetish would be not a simple lie, nor would it be a simple deceiving mechanism hiding intentions behind the appearance of a thing. Although Latour's perspective recognizes how fetish creates actual things that have concrete consequences, this perspective falls into an argument for multiplicity, in which the truth becomes only a play, and as long as you connect your fetish with multiple 'others', it would become more operative, in a sort of postmodernist argument for diversity. What this argument fails to do is to take into account—behind the production of truth—the game of power, the imposition of a 'will' and the notion of the 'whole'.

When Lukács (1971) points to the centrality of the notion of the whole, he is implying that in the production of things—mediated by fetish—therein lies a way of reasoning and an ethical conception of man and society. As we have seen, in his conception, in each part, each singular thing, there is the presence of the historical context and of a philosophical understanding of the world. For him, the mathematical and pragmatic reasoning reduces social relations to things as they exist here and now, thus ignoring their historical formation, which usually is based on blood, violence, or exploitation. Furthermore, the conception of things implies an ethical view of the whole community of man, including an understanding of what is fair, an understanding of how life should be lived and what is dignifying in existence, a division of labour, a conception of the agents in play, a conception of politics and economy, and even an understanding of how we should or should not respect the points of view of others.

Therefore, fetish is not the same as, for instance, the idea of 'method'. Fetish is not simply a procedure, but it is a mystifying procedure, it gives birth to an event—in the sense of Baudiou as we saw in Chap. 3, with ethical consequences—and creates reality. However, it does that with the appearance of a 'naturally existing' thing.

In this sense, fetish is not magic (i.e. where the magician and the public know it is all a trick) because he who acts by means of fetish leads to the belief in that truth. Thus, fetish is eminently a social device, it actualizes an ethics in reality, and it is subjected to struggle between competing arguments.

Therefore, when Mies Van Der Rohe praised 'Less is More', he was actually saying 'if you hide the true artifices of the building process of architecture beneath a clean appearance, you will make people believe that that intriguing and unnatural beauty was achieved with pure procedures, thus they will be strangely seduced by it'. Nevertheless, he might have actually believed he was only doing 'less' and not hiding the 'more', just as the Egyptian priest performing his fetish, in the quotation above from Mary Woods, was acting with the knowledge available and controlled by him, thus making this knowledge real—as much as this knowledge was making him real—through the actual construction of a temple. Similarly, Mies Van Der Rohe hides the constructive pillars of the Barcelona Pavilion by adding a clean chrome cover—what is, at least, one thing 'more'. Additionally, the massive marble walls are in fact hollow spaces. In the unseen void the 'more' is hidden: the effect of a clean massive wall is produced by placing massive blocks in the corners (so that there are no visible joints in the marble), while the middle blocks are in fact thin marble finishing plates covering the elements of support. Thus, perhaps unconsciously, the fetish of 'less' is produced by hiding the 'more'.

When Marx (1990, pp. 163–177) spoke of the fetish of the commodity, he was forging the intellectual weapons of an imminent struggle in a period between the age of revolutions and the age of extremes (as Hobsbawm would term it). Thus, Marx was emphasizing that what the powerful regarded as simple facts were—concretely—the products of the historical development of capitalist society. The aim here is to reveal how intentions are hidden behind the affirmation of certain social facts.[1] Thus, fetish can be understood as a key element in the struggle against the reproduction of society. Furthermore, as in Adorno's quote above, the analysis of fetish is not to deny its reality but to help build the 'consciousness of the objective context of delusion... Its objective goal is to break out of the context from within' (Adorno 2004, p. 406).

In the next section, we will look behind the stage to unmask the tricks of the magicians, but only as a first step to, in the following section, understand how the whole theatre operates, placing active subjects in the stage and passive spectators in comfortable seats. In this way, just as Hamlet's theatre in the theatre revealed the phantasm to be true, we aim to turn fetish against itself and reveal how it is the means of the architectural objectification of architects.

[1] If we return to the example Schopenhauer's cavern, there is a 'will' that connects the shadows both inside and outside the cavern. And this will is not simply a universal spirit but the concrete manifestation of an ethical approach to society towards the construction of social objects.

5.2 Techniques of Phenomena: Conceptualizing Fetish in Space

We have seen how architecture is not a simple thing. It is reified social relations, and furthermore, this reification frames subjectivities in cities. But how do different subjects achieve this reification?

A Dancing Table

Marx (1990, pp. 163–164) started his conclusion regarding the fetish of the commodity by analysing the secret behind a simple table. He urged the reader to turn that 'sensuous' form (with its four legs) upside down in order to see how it 'evolves out of its wooden brain grotesque ideas, far more wonderful than if it were to begin dancing of its own free will'. It is not only because behind its forms, still made of wood, the table also becomes an object to be exchanged, thus creating interactions between subjects in the 'market'. Furthermore, that 'thing' the table 'is' is not only what architects would normally make of it: lines, planes, shapes, and styles. It is not even what more sensible architects would envision as social constructions encircling a cultural habit of sharing food. This metaphorical dancing table 'rather transforms every product of labour into a social hieroglyphic. Later on, men try to decipher the hieroglyphic, to get behind the secret of their own social product' (Marx 1990, p. 167). In this dancing table, Marx finds the hidden means through which it was produced: the social knowledge involved, the science of materials, the symbolic value it carries, the mobilization of desire, the logic of commodities, the division of labour and, furthermore, subjects enclosed in a chain of actions. In other words, he finds the hidden forces reproducing social relations by means of an object of design—he finds, namely, the fetish.

In this sense, the fetish of this dancing table cannot be understood either under the sign of a simple verb or as a simple noun. The mysterious character of the fetishism of commodities in Marx has the ambiguous meaning that it is both a (social) action and the result of this (social) action. It has the ambiguous property of being produced by social relations and of being productive of social relations. It has the ambiguous property of being believed to (or seeming to) have powers that it does not have and at the same time constituting forces that could not exist without it. It has the property of being simultaneously 'subjective' and 'objective'.

It is in this sense that we will use the word fetish as a present participle, in the original Latin sense of being at the same time an adjective, a noun, and a verb. In this sense, the fetish in architecture can be seen as the characteristic in an object or as the constitutive forces of this object. Therefore, fetish can be understood as a verb in the original Latin *partem capit* (literally meaning 'an action' that 'takes part'—a present action). This does not intend to add imprecision to the concept. Rather, it draws upon the imprecision that makes fetishism possible and causes past reified social relations to be accepted as natural phenomena. Only in this way can we understand how the table is not a table that can dance; rather, it is a dancing table in its very stillness.

Thus, the aim of this chapter is to understand how architectural operations inhabits things, continuing to act in their very stillness. We aim to understand how architecture mobilizes mediated social processes in the form of immediate truthful things. We aim to understand how architecture transforms 'things into persons, and persons into things' (Marx 1990, p. 209). In this sense, we will need to unveil the black box of the non-trivial machinery of architecture, we will need to unveil the enigmatic operations behind the production of architectural things, and we will need to unfold the methods implied in the production of 'architectural truths'. Only after deciphering this hieroglyphic enigma, we will be able to recognize, later on in Chap. 6, how architects ultimately subject themselves to the fetishes of the discipline (namely, the objectification of architects themselves by means of shared phantasies).

Phenomena Techniques

However, the fundamental question here is how does fetish come to inhabit things? In Latour's (1986) work previously mentioned, he further investigated how abstractions would acquire life when they were reified into technical apparatuses. When a scientist uses an apparatus to observe phenomena, what he sees on the other side of the 'black box' of the apparatus is framed by the past theories and hypothesis that produced that 'black box'. Thus, he argues, the phenomena this scientist sees only existed through the mediation of the machine, and the machine only existed through the past labour reified in it (the theories inscribed in this material basis). In his words,

> The central importance of this material arrangement is that none of the phenomena 'about which' participants talk could exist without it. Without a bioassay, for example, a substance could not be said to exist. The bioassay is not merely a means of obtaining some independently given entity; the bioassay constitutes the construction of the substance. Similarly, a substance could not be said to exist without fractionating columns (Photograph 7) since a fraction only exists by virtue of the process of discrimination. Likewise, the spectrum produced by a nuclear magnetic resonance (NMR) spectrometer (Photograph 8) would not exist but for the spectrometer. It is not simply that phenomena *depend on* certain material instrumentation; rather, the phenomena *are thoroughly constituted by* the material setting of the laboratory. The artificial reality, which participants describe in terms of an objective entity, has in fact been constructed by the use of inscription devices. Such a reality, which Bachelard [...] terms the 'phenomenotechnique', takes on the appearance of a phenomenon by virtue of its construction through material techniques (Latour and Woolgar 1986, p. 64).

It is in this sense that he explains the 'black box' of laboratory equipment using Bachelard's idea that these scientific apparatuses are 'reified theories': 'When another member handles the NMR spectrometer (…) to check the purity of his compounds, he is utilising spin theory and the outcome of some twenty years of basic physics research' (Latour and Woolgar 1986, p. 66). The ideas inscribed and configured on the machine were based on arguments and theories, and these were the results of discussions at conferences and disputes in journals and articles, until they were finally accepted as facts. Thus, the 'so-called material elements of the laboratory are based upon the reified outcomes of past controversies' (Latour and Woolgar 1986, p. 87).

To create a theory of photography, Vilém Flusser (1985) developed a 'philosophy of the black box'. Flusser (1985, p. 43) argued that it is the photographic camera that performs the operation of transforming reality into codified signals of visual communication, and the photographer is manoeuvred by the few potentialities inscribed in the apparatus. Therefore, the photographer actually looks inside the apparatus rather than outside, thus 'revealing' rather than creating. However, for Cabral Filho and Baltazar (2010), analysing Flusser's theory in the realm of art and technology, it is not a matter of destroying the 'magic' of the black box or of making its devices predicable and dull but rather opening its internal mechanisms for potential interactivity. A music box would have interactivity, but only repetition; a piano does not reveal its content but allows creation and interactivity, but in the same way as a photographic camera, it does so within a framed realm of possibilities. The challenge is to open the inner realms of the devices for interactivity and creation.

In the field of architecture, the emergence of apparatuses such as CAD, renderers, and 3D software are increasingly 'entailing' the production of architecture, in the same sense that a refrigerator 'entails' a power source (to use an example from Taylor 2010, p. 44), thus transforming architects into operators of machined global styles who share content that is capitalized by companies providing access to this content. Furthermore, one could agree with a Mark Cousins talk in 2016, when he asserted that whereas previously, the artist had a symbiotic relation with the 'brush', and the architect with the 'pencil', today the ultimate apparatus of architectural production is the 'office'. Arguably, one that enables a 'genius' to seduce followers and to channel the work of a legion of workers. With a high-profile office, an architect can expropriate the work of hundreds of others. However, this non-trivial machinery only works if architects desire to be part of it.

Deleuze has famously asked: how do people come to desire their own oppression? For Zizek (2014), the answer of what we desire lies in fantasies: the narratives a subject creates to build a logical chain of causalities that assures his desires as unconditional. In architecture, once the fetish of the genius architect has collapsed, these phantasies start to emerge raw and naked. This is the topic of the next section.

5.3 Abstract Machines and Fantasies: Princes and Utopia

Something quite different from the last century's foundations for spreading the ideals of dead architects. Once upon a time, Le Corbusier, Buckminster Fuller, Frank Lloyd Wright, Mies Van der Rohe, Lina Bo Bardi, and practically all major figures of modern architecture left foundations that are responsible for spreading their convictions, developing research, supporting publications, funding disciples and reproducing the architects' ideals. This was a mechanism for safeguarding a prominent and increasingly powerful position in the field of architecture.

In 2012, when Oscar Niemeyer died at the age of 105, he was still 'designing' a vast number of buildings around the world. A few days after his funeral, his grandson, the architectural director at Niemeyer's office, declared that the firm would finish the projects already started, setting an end to 78 years of architectural practice, even if one could not measure the size of Niemeyer's contribution towards the end of his life. Nevertheless, Niemeyer remained very talkative and engaging in his lectures, always surrounded by admirers astonished by his unceasing commitment to designing. As expected, Niemeyer left a foundation to protect his legacy, even if his grandchildren are now in a fierce legal battle over his heritage.

The death of Zaha Hadid, in early 2016, led to a very different turn of events. At the time, Zaha Hadid Architects (ZHA) had offices in London, Beijing and Hong Kong. These offices had already developed more than 950 projects, and the firm was about to open a new office in New York, with additional plans for offices in Dubai and Mexico. Some weeks after Hadid's death, finding the world of architecture in mourning, Patrik Schumacher, who had become a partner at ZHA in 2002, declared to the NY Times that ZHA would continue designing (Erlanger 2016). According to Schumacher, Hadid had set a precedent; she had imbued the practice of architecture with a 'new repertoire' and a new 'spirit'. On this basis, he could ensure that the firm's 400 staff members could confidently continue her vision and research (see Erlanger 2016). Thus, Zaha Hadid lives:

> any star in architecture has been born in the discipline itself, and emerges through schools, competitions and colleagues. (…) We want to tell the world that we're still a viable, vibrant address for major work of cultural importance. (…) My ambition is to become more visible as a leader of the field to clients (…) This star signature is a relatively new phenomenon (…) We feel very confident that we will carry on and go forward with her vision and her legacy and the experimental research she established in the office (Patrik Schumaker *in* Erlanger 2016).

Arguably, these spirited architects controlling a design practice by means of supernatural means is a new phenomenon in architecture. However, this is not a new phenomenon in other fields, such as the fashion industry, where many designer names have continued after the deaths of the founders.[2] Nevertheless, in architecture, the image of the 'gifted architect' determining the structure of the field with

[2] Consider, for example, Coco Chanel, Gucci, Versace, Prada, and Benetton. Except for the first, these are family names, and the companies continued after the founders' deaths, with family members playing different symbolic roles. For instance, when Christian Dior died in 1957, some considered closing the company (see http://www.fundinguniverse.com/company-histories/christian-dior-s-a-history/). Nevertheless, since its beginning, the company had been financed and controlled by the industrial complex of Marcel Boussac, and since the first year, Pierre Cardin had been the head of the 80 employees' creative workshop. Thus, continuing under the artistic direction of Yves Saint-Laurent was a natural move for a business born as a corporation. Furthermore, by 1950, the company started a licensing programme, putting its famous name on dozens of products produced by a wide range of companies worldwide. Although this scheme created controversy among colleagues in high-fashion, who wondered about the pedigree of the inventions, this would

his pen demanded a quite different discourse, and notions of brand used to be quite extraneous.[3]

Another paradox seems inevitable. The history of architecture has traditionally remained distant from the logic of the market, developing its own distinctive qualities apart from market value (Bourdieu 1996a) and building a parallel system of valorization. In this system, biographical narratives have been the most important reference for measuring the greatness of a 'genius'. With few exceptions, ever since the first history of painters by Giorgio Vasari in 1550, the history of art has been depicted as a collection of biographies that chronicle the 'succession of inventions' delivered by each individual artist along a linear timeline, sometimes with the addition of contextual information for background. Could the meaning of 'genius' now return to its original Latin sense of 'spirit', thus animating a brand's industrial business? Or has 'genius' always been part of the fetish of the field?

As Schumacher demonstrates in his first pronouncement after Zaha Hadid's death, in a Facebook post entitled 'Zaha's Incredible Moves', he is very resolute in reinforcing Hadid's genius contribution to the field: 'The bounds of architectural possibilities had shifted. Architectural design gained a whole new dimension'. He is even more direct in the following passage: 'Zaha had indeed delivered an unprecedented expansion of the discipline's repertoire'; and finally, he writes of the 'wholly original and empowering "discoveries" that Zaha gifted to our discipline'.[4]

Thus, the phenomenon of the architectural 'star' introduces new elements, allowing the emergence of a phantasmagorical era in architecture in which the magic produced by these architects does not even require that they be alive. Some new questions could emerge: would it be possible for a well-trained and highly tuned team to continue not only the legacy but also the inventiveness associated with a dead architect? And more radically, could we, for instance, resuscitate Le Corbusier?

This question reveals a conflict in the internal logic of the field; it also reveals the intensification of a contradiction in the discourse on how architecture is produced. So far, the mainstream history, as documented by the academy and the offices, persists with the rhetoric of a great figure: the great (wo)man who had a vision and a set of skills, and who, with his/her hand and pen (or computer), changed the world. We now see a veiled move, intensifying the idea that an architect is not a normal human acting in flesh and bone—a normal human who will one day die. They have been

---

soon become a model for the rest of the industry. Since its start, the Dior brand has been behind a wider chain of creative productions and factories.

[3] Now, Rem Koolhaas has openly advised the team of ZHA to follow in the steps of fashion houses; see http://www.dezeen.com/2016/05/14/zaha-hadid-architects-should-follow-example-of-mcqueen-says-rem-koolhaas/.

[4] For the 'Facebook post', please refer to https://www.facebook.com/patrik.schumacher.10/posts/10208103116688623. It is remarkable enough to notice how a drawing of a student exercise can be used to corroborate the narrative of 'inventiveness' and 'innovation' in the field even though it mimics experiments—including architectonic ones—carried by Van Doesburg, Gerrit Rietveld, and El Lissitzky (cf. Frampton 2000, pp. 157–178) more than half a century before. Thus, to build an illusionist narrative, a chain of causalities is reshuffled by manipulative discourses and reproduced by the main institutions in the field.

replaced by a new spirited genius who, rather than producing architecture directly with their hands and brain (as a living worker), is a type of device that manages complex and abstract processes and coordinates the actions of hundreds of other architects and employees. This is a step closer to the image of a prince(ss) who is nothing more than a representation of collective work. Thus, Zaha Hadid shall live.

This condition only reveals the gap between the actual means of production of architecture and the narratives of the field. While the narrative is that Zaha Hadid's personal cleverness created a whole new field,[5] this narrative is nothing more than a representation of the work of 400 employees and the (human and financial) capital necessary to maintain her office from 1979 until her first built commission in 1993. It is important to note that this myth of ingenuity is widespread in today's culture. As we saw in Mariana Mazzucato (2011) example of the iPhone invention, which major innovations have come from state-funded research, i.e. they have been collectively produced. Then, such innovations have simply been appropriated by companies and produced by thousands of workers overseas. In this sense, Mazzucato's argument reinforces the idea that production is the result of a social and collective process—a 'general social knowledge' in the words of Marx (1973 [1857])—that we shall examine below.

Fantasies Capturing Collective Subjectivities

Arguably, the movement known as 'Autonomism' has investigated production through a renewed analysis of Marx's (1973 [1857]) account of the 'general intellect' in the 'Fragments on Machines' in his notes called the *Grundrisse*. There, Marx proposes that the actual force of production is the general knowledge that results from society's functioning as a complete ensemble. Nevertheless, this knowledge is increasingly objectified into machines, which, in turn, function as devices to capture more of society's collective productive forces. Thus, machines function as means of control and expropriation: 'In machinery, knowledge appears as alien, external to him [the labourer]; and living labour [appears as] subsumed under self-activating objectified labour' (Marx 1973 [1857], p. 695). This passage in Marx has been very fruitful for contemporary critical theory because it envisions how the actual production of innovation, science, and technological development occurs under the development of mechanisms to frame and capture social work.[6]

[5] Certainly, there is a question of belief at play here. Did the Pharaoh really believe he was God? Did Zaha Hadid really believe she was a genius? How come both we and they can believe in this kind of social mystification? There is a mode of visibility at play beyond personal opinions, a social fantasy that frames the possibilities of what we can—and cannot—believe.

[6] The fundamental contradiction between the (neo)classical economy and a (neo)Marxist political economy is thus the theory of 'Labour' value. In Marx, the value of a product comes from labour. Indeed, for Adam Smith and David Ricardo, that was also the case. However, mathematically, this is incompatible with any ethical justification of capitalism because value must be extracted from

For Paolo Virno (2001), the dialectical nature of Marx's materialism could not be seen as today's 'Marxism', as for Virno, Marx reveals how the material conditions of production are objectified abstract scientific knowledge fixed into capital (one could say 'past labour'), thus revealing an 'inter-subjective foundation' in any labour praxis.[7] For Lazzarato (2014, p. 31), this radically changes the search for a subject of history, as this condition is neither a representation nor a consciousness but rather a mechanistic entanglement of molecular parts. In this realm, there are no individual subjects being dominated (as in personal enslavement), rather, there is diagrammatic management of all workers in society.[8]

In addition, David Harvey (2010, p. 40) stretches the emphases of Marx's *Capital* into the 'roles' people play in the market system. In this system, social relations are presented as an exchange of things (Harvey 2010, p. 41). In addition, even if one might have ethical and moral principles when dealing with people face to face in the buying of a commodity in the market, these relations appear as relations between things (commodity-money) and therefore as inevitable facts. For instance, at the moment you buy bread, you are reinforcing the system as much as your retirement fund is managing assets in the global market. For this reason, Harvey (2010, p. 47) says this creates an unavoidable condition, which imposes specific 'roles' because 'the characters who appear on the economic stage are merely personifications of economic relations'. Therefore, people will, even unwillingly, become 'the bearers' of capitalistic social relations.[9] In this sense, it is not a matter of a 'bad' architect

---

the workers by the capitalist. This sheds light on how neo-classical economist, such as Michio Morishima (1977)—which demonstrates the incompatibility of micro-economy and labour theory of value—are forced to ignore ('abandoning') concrete labour to make their theory coherent.

[7] For Virno (2001), the fundamental contradiction between knowledge and product did not result in the breakdown of capitalism due to new forms of domination in post-Fordist societies. For Virno, the mass intellectuality that is not objectified in machines is later captured by a mass control of communication and sharing. Furthermore, a politics of affects and cynicism makes possible a widespread pseudo-solidarity with those suffering, and at the same time ferocious forms of competition. As Jason Reads puts it: 'Competition is a paradoxical form of individuation in that it produces individuals who are all the more alike in that they see themselves as absolutely opposed to each other, locked into bitter struggle' (Read 2010, p. 130).

[8] Some critics are correct about the contemporary neoliberal emphasis on creative and immaterial labour as a more radical form of expropriation, but they presuppose that critically approaching a topic corresponds to a naïve acceptance of the process, thus equating praise (Florida 2002) and critique (Lazzarato 1996) (see Brouillette 2009). Nevertheless, a careful reading of Lazzarato (2014), searching for context rather than fragmented contradictions, could not avoid the fact that Lazzarato is well aware of the paradoxes involved (as Marx was in his fragments), aiming to develop an account of how labour is increasingly being reified by abstract machines, thus, no being is a naïve safe harbour for freedom. Therefore, it is fair to say Lazzarato is not Florida. As it relates to our study in this moment, it will suffice it to say that the autonomists and such are trying to develop, as criticism, their own version of a Gramscian subject, be it the 'multitude' or the 'enlightened'. Rather than asking if the worker is the 'soul', 'stolen', or 'colonised', we aim to investigate how, besides being produced, these subjects are theoretical devices and how these subjectivities are socially produced and composed transversally and meta-individually.

[9] Harvey elucidates this point with an example: 'recognizing that individuals can and do often occupy several different roles, even deeply contradictory positions (as when, in our time, a worker

enslaving another but rather of a whole field of practice working as a collective apparatus reifying subjectivities, which is producing new object-subjects of fetish.

Thus, Marx's fragment on machines and his considerations of the production of social relations in the form of fetishism led many authors to investigate how subjectivity is produced in contemporary society rather than to investigate how subjects could free themselves, find their supposed true natures and be autonomous individuals acting in the historical process (Guattari 1995; Lazzarato 2014; Read 2010; Spencer 2012).

For Jason Read (2010, p. 155), the expression 'production of subjectivity' has a double meaning: as something 'productive' and as something 'produced'. Subjectivity is historically produced by multiple processes of individualization in physical, biological, collective, psychic, linguistic, and cultural sensibilities and through power struggles. At the same time, it is productive of social relations that impact the possibilities of action in society. Read rescues this idea from Marx's Capital:

> the special productive power of the combined working day, is under all circumstances, the social productive power of labour, or the productive power of social labour. This power arises from cooperation itself. When the worker co-operates in a planned way with others, he strips off the fetters of his individuality, and develops the capabilities of this species (Marx 1990, p. 447).

Subjects are always subjects in a collective, therefore, a differentially articulated part of a whole, and the individuations of a metastable field of transindividuality. This means that a subjectivity is formed by a priori elements (language, culture, structure, social expectations, and so on) 'externalised in machines and internalised in concepts, habits, and ways of thinking' (Read 2010, pp. 118–119). The political problem emerges regarding how these machines operate not as atoms but transversally, capturing the collective intellect formed by society.

In the contemporary mode of production, there are many examples of such processes. The biggest new businesses are only abstract machineries, platforms that capture not only the labour of others but also their everyday lifestyle. Today, Facebook is the biggest platform for sharing information, which it does without producing any content, only capturing a series of 'interconnected' users gladly producing content as a form of leisure, without recognizing it as production. Uber might be considered the biggest transportation company in our 'smart' times;

---

has a pension fund invested in the stock market). This focus on roles rather than individuals is as perfectly legitimate as if we were analysing the relations between drivers and pedestrians in the streets of Manhattan: most of us have taken on both roles and adapt our behaviours accordingly' (Harvey 2010, p. 48). The implication, he explains, is 'that our social relation to the labouring activities of others is disguised in the relationships between things. You cannot, for example, figure out in the supermarket whether the lettuce has been produced by happy labourers, miserable labourers, slave labourers, wage labourers, or some self-employed peasant. The lettuces are mute, as it were, as to how they were produced and who produced them [...] It is all very well to insist on "good" face-to-face relations and to be helpful to one's neighbour, but what is the point of that if we are totally indifferent to all those whom we do not know and can never know, but who plays a vital role in providing us with our daily bread?' (Harvey 2010, p. 40).

without owning or maintaining any vehicles, it transfers all the risks of the business to its workers while it also exploits the socially produced infrastructure of the city without paying taxes for it. Similarly, Airbnb offers the largest variety of lodging without owning any of the properties, but it still creates new subjectivities of trendy travellers and kind hostesses based on people gladly providing undervalued goods and/or services in exchange for a ‘social experience’. This goes far beyond Le Corbusier’s trivial house-machine or the pathetic concrete life in Koolhaas’ trivial machinist house depicted in the film ‘Houselife’. Closer to the new non-trivial machines are initiatives at the MIT Media lab, which are creating automatons that reproduce ideological systems through the deployment of algorithmic fetishes in architecture (see Shvartzberg 2015).

Fantasies, in this context, refer to the narratives that mobilize the desires of users, designers, and stakeholders, shaping the illusions of the field and preventing critical reflexivity. The main phantasies of architecture—'utopias' and ‘princes’—operate as blindly accepted narratives that define architectural practice. These narratives facilitate a continuous process of appropriation of authorship, leading to what may be termed the ‘surplus of creation’.

It is crucial to distinguish between fantasy and phantasy. A fantasy is a conscious imaginary construct, an image or representation of a non-existent object that can be desired or despised, while a phantasy operates unconsciously, structuring methods, senses, perceptions, and reasoning. Phantasies exist a priori, shaping subjective senses in ways that architects often remain unaware of, as they believe their methods to be impartial and universally valid.

According to Slavoj Žižek’s interpretation of Lacan, phantasies are conscious narratives developed to justify and mobilize unconscious desires. This understanding underpins the reproduction of architectural practice and the estrangement of architects from the means of their production. Phantasies provide a framework through which subjects navigate their desires, forming the foundation of architectural ideologies. For Žižek,

> The first thing to note is that fantasy does not simply realize a desire in a hallucinatory way: rather, its function is similar to that of Kantian ‘transcendental schematism’: a fantasy constitutes our desire, provides its coordinates; that is, it literally ‘teaches us how to desire’. (…) To put it in somewhat simplified terms: fantasy does not mean that when I desire a strawberry cake and cannot get it in reality, I fantasize about eating it; the problem is, rather: how do I know that I desire a strawberry cake in the first place? This is what fantasy tells me (Žižek 1997, p. 7).

As subjective senses assume an actualized individuation in the subject through the production of phantasies (a narrative that provides linear causal links to justify one’s desire), Žižek argues that desire does not create fantasy; instead, phantasy produces desire. For Žižek,

> Desire emerges when drive gets caught in the cobweb of Law/prohibition (…) and fantasy is the narrative of this primordial loss, since it stages the process of this renunciation (…) In this precise sense, fantasy is the very screen that separates desire from drive*, it tells the story which allows the subject to (mis) perceive the void around which drive circulates as

> the primordial loss constitutive of desire. In other words, fantasy provides a rationale for the inherent deadlock of desire (Žižek 1997, p. 43).

This perspective challenges the substantialist notion of 'drive' in Freud's theory. Žižek's argument retains drive as a foundational element, conceiving phantasy as either a mask for a deeper substance or a layered construct of a void. However, if we follow Vygotsky and Simondon's theory of transindividuality, phantasy is better understood as the production of subjective senses, actively shaping the architect's perception of reality. Phantasies thus become instrumental in defining how architects relate to their conditions of production.

Lev Vygotsky's concept of 'subjective sense' offers a key perspective on how phantasies shape subjectivities. González Rey (2007) expands on Vygotsky's insights, arguing that subjective sense is formed through lived experiences, historical contexts, and the interplay of symbolic and emotional factors. Unlike Freudian psychoanalysis, which universalizes the subject through an essential libido energy, Vygotsky's dialectical psychology emphasizes the integration of cognition and affect within specific socio-cultural settings.

Architectural phantasies reproduce specific ways of thinking and valuing architecture, ultimately disciplining subjectivity. As Althusser (2000) argues, the narcissistic view of the world functions as a 'doubled mirror', not merely reflecting reality but projecting specific subjectivities as if they were objective truths. This projection traps architects in reified narratives, sustaining the structures that dictate their practices.

Deleuze famously asked: how do people come to desire their own oppression? According to Žižek (2014), the answer lies in phantasies: the narratives subjects create to build a logical chain of causalities that justify their desires as unconditional. In architecture, once the fetish of the 'genius architect' collapses, these phantasies emerge raw and exposed, revealing their ideological function.

Through architectural phantasies, ideologies are internalized, influencing units of subjective senses and shaping the cognitive and affective frameworks of architectural subjectivity. Phantasies reproduce disciplined practices, maintaining the status quo by ensuring that architects remain unaware of the structures shaping their perceptions and decisions.

Despite their reifying function, architectural phantasies can be critically examined and traversed. Architectural subjectivity is, therefore, the historical inheritance of a culture shaped by past subjects within specific ethical and social contexts. By unveiling the phantastic truths embedded in architectural practice, we can move towards a more critical and reflexive understanding of the discipline. Untangling the conditions shaping the desires of architects might allow to reclaim agency over their own production.

The Prince and Utopia as Phantasies

Althusser (2000) understands Machiavelli's prince as a device that manages fortuna—the unpredictable contingency of context—to capture public will through virtue, the wise and rational orchestration of collective desire, achieved, we could say, through phantasies. The prince is not power himself; instead, he is an effective public image concerned solely with producing a certain truth rather than truth itself. This image supports a mystical narrative of social affairs, capturing expectations to build a linear causal chain that reinforces the prince's power. In this sense, the prince is not a substance or a structure but a guiding image whose power lies in its ability to direct operations. Only by means of this fetish does the prince become the subject of transformation.

Applying the concept of the prince to architecture better captures how architecture interpellates the subjectivity of architects, as it accounts for the politics of fetish and the mobilization of transindividual desires. While critiques of 'star architects' attempt to challenge the myth of the genius architect, they often reinforce it instead. The term 'star' implies inherent qualities or magnetic influence, but like actors, these figures are mythological constructs—representations masking the collective means of production. A close examination of actors would reveal their dependence on marketing, professional agents, class hierarchy, and social networks. Similarly, comparing architects to film directors or producers is flawed, as they too are merely effective images concealing collective labour.

Tafuri's (2006) made significant advances in contextualizing architectural production, but his critique ultimately reinforces the interpellation of the 'genius' by focusing on individual architects rather than the broader processes of subjectivation and the production of phantasies. In Simondon's terms, Tafuri critiques already individuated architects as though they were fundamental substances, rather than examining the individuation process itself. His argument moves from asserting that the 'base' determines architecture as an automatic effect to focusing on the effect (individual architects) rather than the process (the formation of subjectivities and phantasies). The prince architects he critiques are just images capturing what is, in reality, a transubjective and collective work.

For example, although Tafuri and Dal Co (1976, p. 140) noted that Frank Lloyd Wright's decisive stylistic shift was made possible through his wife's fortune, he fails to address how this project was produced. Wright used this capital to create the 'Taliesin Fellowship', a messianic school in the desert, through which he appropriated the work of numerous collaborators and apprentices as his own (Friedland and Zellman 2007). Wright's style was not solely his; it was the product of collective labour, appropriated by the myth of his genius. Despite the financial strain of Broadacre City, it solidified him as a symbol of US architecture—an intentional move on his part, as he sought to become the icon of an American architectural style to be propagated globally. Architecture thus becomes a means of capturing and reproducing collective subjectivities. The crucial issue is not merely recognizing that Broadacre City reflects American society but understanding how these phantasies—both the prince and the utopia—reproduce reality by colonizing subjectivities.

In our contemporary society of spectacle and immaterial toil, the 'princes' have imploded upon themselves. Once the architectural office was an elite stronghold dividing intellectual and manual labour; now, it is the new sweatshop. Bjarke Ingels recounts his experience at Rem Koolhaas's OMA office, where rising in rank required accumulating 'more and more sorrows' by taking on increasing responsibility to 'crank out cool stuff' (Parker 2012). He describes an environment of yelling, thrown office supplies, and constant stress induced by negative reinforcement. Eventually, he left to establish his own office—only to replicate the same exploitative structure, replacing negative reinforcement with behaviourist 'positive reinforcement'.

A critical approach to Machiavelli's prince must therefore examine the process of becoming an architect. Althusser's interpretation of Machiavelli provides a framework for systematizing this fetishist practice in critical terms, incorporating the political dimension in the production of accepted collective narratives—socially shared causal chains. The ontogenesis of individuated architectural princes exposes a preindividual field that not only frames possible individuations but also constructs structural inequalities, reproducing reification and fetish. Understanding this process is crucial to traversing the phantasies that sustain architectural subjectivity.

Our exploration of utopian phantasies in architecture should begin with Mannheim's inversion of the correlation between 'reasonable-utopia' and 'possible-impossible' (Mannheim 2000, pp. 173–175). He argues that interpreting social reality as 'natural' ignores the historical processes that produced it. Ideology, by treating abstract ideas as if they were already realized, is itself 'u-topic'—out of place. For instance, ideological notions such as 'all men are born free' assume a realized state that may never materialize within existing social structures.

If ideology facilitates 'the conversion of things into persons and the conversion of persons into things' (Marx 1990, p. 209), utopia, in architecture, serves to justify and reproduce this order. It functions on two levels: as a justification of existing structures and as a means of their reproduction. Classical utopian thought focuses on justification, whereas pseudo anti-utopian approaches obscure the mechanisms of reproduction. This distinction brings us to the difference between fantasy (conscious) and phantasy (unconscious), which helps to clarify the role of utopia in architectural discourse.

On the one hand, utopia operates in architecture as fantasy, providing a narrative that justifies an architect's propositions, mobilizes desire, and reinforces existing social relations through the production of an image. This conscious articulation serves as a legitimizing story, making utopia a 'presented' object that the architect desires. Fantasy, in this sense, is an explicit narrative.

On the other hand, utopia also functions as an 'absent' object of desire—phantasy—by embedding itself in the discipline of architecture as a method of social fetish. Architects, unaware of these embedded processes, perceive their methods as impartial technical approaches, when in reality, these methods construct preindividual desires into generalizations and systemic requirements. Here, utopia is phantasy: an unconscious structure shaping architectural thought and practice.

Traversing Phantasies

Hamlet, the prince of a rotten kingdom in Shakespeare's play, was informed by a phantasm that he was living in a net of lies and simulations. Thus, he became melancholic,[10] trapped between two worlds: one of the pompous lies of the court, which made his uncle a king; and the other the phantasmagorical truth, revealing all to be a game of appearances. To test whether the 'accepted courtly phantasies' or the 'phantasmagoria awareness' was the truth, he staged a play inside the play, a theatre inside the theatre, which revealed the lies of his uncle-king.

Simply rejecting utopia is futile; instead, we must recognize its inescapable phantastic nature. To move beyond these unconscious constructs, we must expose the artificiality of our desires and the fetishist nature of their production. This effort reveals the political dimension of phantasies, showing how all proposals are founded on ideological representations of reality. In this sense, Jameson's (2004) ultimate argument is that the value of 'utopia is somehow negative':

> …and that it is most authentic when we cannot imagine it. Its function lies not in helping us to imagine a better future but rather in demonstrating our utter incapacity to imagine such a future—our imprisonment in a non-utopian present without historicity or futurity—so as to reveal the ideological closure of the system in which we are somehow trapped and confined (Jameson 2004, p. 45).

For this reason, Jameson asserts that 'each of these utopias is a fantasy, and has precisely the value of a fantasy' (Jameson 2004, p. 50) and asks the following question: 'Is this to say that we can form no substantive or positive picture of utopia, short of embracing all the multiple contradictory pictures that coexist in our collective social unconscious?' (Jameson 2004, p. 51). Answering this question presents a striking challenge for architectural theory, under the weight of its dearest myths: the prince, the utopia, the facts, and the architectural object itself—all fantasies.

On the one hand, the 'prince' is a powerful political device, as Machiavelli recognized more than 500 years ago. Prince representations are not simply lies, they have true consequences; their effectiveness lies in their ability to structure reality, shaping collective subjectivities and colonizing existing power relations. This myth, in turn, is grounded in another: the myth of the individual. Architecture clings to a narcissistic epistemology, one that remains trapped within its own reflection—the architect's self-mythologizing gaze. On the other hand, utopia, a cherished concept in architecture, is also a myth: a myth of a myth. Utopia does not transcend the *topos*; rather, it rearranges the status quo. But its power resides precisely in being

[10] Here we extend Vygotsky's (n.d.-a [1925]) long analysis of the play, to counter the commonly accepted interpretations that Hamlet was pretending to be mad.

pure desire, capable of reproducing narratives. Yet, escaping phantasy is not an option; the task is to traverse it, making its mechanisms visible and its ideological functions explicit. If that is the case, all mainstream histories of architecture become a trip through compendiums of collective phantasies—though not something entirely without its value.

Our ultimate aim is not to find the positive truth behind the fake appearance of these phantasies; instead, we aim to unveil how these phantasies have concrete implications in reproducing subjectivities. This idea comes from Lacan, who argues that a subject cannot escape phantasy but can recognize the elements that constitute his desire and its representation in objects, thus enabling the subject to act reflexively towards his impulses.

Žižek argues that Lacan is a key reference to discuss the 'traversing of phantasies'. For Žižek, the subject can avoid the confusion of 'what he is' and 'what he desires' not only by recognizing the phantastic frames setting these desires in motion (1997, p. 62) but also by closing the gap between 'imposed' choices and 'framed' choices (Žižek 1997, p. 40). Vanheule (2011, p. 8) further explains the importance of 'traversing the fantasy' for Lacan:

> Traversing the fantasy, in its turn, means that through ongoing free association, the positions the analysant has been occupying in relation to the object 'a' [the unattainable object of desire] are mapped with the aim of enabling different positions toward the Other's desire and toward the object 'a'. The fantasy is traversed with the aim of breaking the repetitive cycle of incarnating the same position in relation to the Other and to the object 'a'.

In this sense, our aim is to enable subjects to break the cycle of repetition and reproduction of its subjectivities. However, rather than understanding the 'Other' as the (absent) 'structure' of language or any other deeper 'dark matter' in the individual (as does the first, the middle, and the last Lacan, respectively, representing it in a mirror, in a semiotic double mirror, and in a Moebius individual inner substance), we understand the process of instituting these phantasies as historical and collective operations, i.e. as the social production of collective subjective senses.

Ideally, this traversing of phantasies could enable a subject to reposition himself in the field that reproduces his own desires. Changing the position of the subject also changes the projection of the map (just as the repositioning of the camera changes the picture). Thus, who repositions himself also produces a new map. This does not eliminate the projection distortion; rather, it only makes it identifiable, traversable, and available to reflexivity.

In this chapter, we aimed to map how the urban is framed by subjectivities and how that in turn further reproduces the subjectivities of those who design the urban space. In the next chapters, we aim to trace emergent routes for oppositional practices and establish some principles to overcome the myth of the princes. To do so, we will analyse how counter-hegemonic practices are critically addressing these issues, so in the last chapter we can try to reshape our understanding of how the city could be transformed—its narrative of causal change—and how we can reshape the conception of the 'subject' who creates the architecture of the city.

Chapter 6
For Deconstructing Neoliberal Objectification

Abstract This chapter develops a participant observation of design practices and research aimed at overcoming neoliberal practices. It formulates key contradictions in these practices by tracing the theoretical framework developed so far. In addition, it explores key concepts that might feature these practices, such as the hacking of objectified social relations, living projects seen as processes rather than as the production of things, commoning practices against urban enclosures and micro-utopias seen as retroactive, concrete, or immanent versions of the classical utopia. Furthermore, it assesses the extent to which these practices are capable of developing alternative design propositions.

Keywords Counter-hegemonic practices · Micro-utopias · Hacking · Collaboration · Architecture as process · Deconstructing spatial objects · Subjectivity · Spatial agency · Commoning · Live projects

6.1 Live Contradictions in Architecture

Practices in Live Laboratories

The current transformation of public spaces in contemporary cities does not occur without struggles and forms of resistance, both outside and inside the discipline of architecture and urban design. We can generally organize them into two groups, one occupying space and the other developing critical interventions in the production of space.

On the one hand, recent economic situations have created a series of social movements that are directly engaged with the appropriation of space. The most notable of these movements started with the Occupy Movement, which aimed to regain control of society by occupying space (but many other urban conflicts emerged, both symbolic and physical). As Smith and Low argue, no revolution can occur without taking over the space of the city (Smith and Low 2006). In many situations, a practical form of 'surrealism' was born in the streets. Protesters would transform everyday

C. V. De Lima Amaral, *The Urban Reproduction of Subjectivities*,
SpringerBriefs in Geography, https://doi.org/10.1007/978-3-032-01465-8_6

objects and urban furniture into occupying weapons, subverting the spatial rules of such spaces (Singer 2014). In these strategies, the experience of urban art and social media collaboration was fundamental in spreading ready-made strategies.

On the other hand, a series of experiences have emerged to confront the architectural field's traditional boundaries. Architects have engaged in alternative forms of spatial interventions in the city. This dialectic of a soft mode of control, urban struggles, and architectural practice has developed an entirely new set of social intervention strategies. As an example, a series of small-scale experiences were conducted by architectural groups such as DK_CM, Assemble, and Atelier Urban Nomads.

David Knight and Cristina Monteiro from the office DK_CM developed the project 'Building Rights' that aims to develop a socially organized systemic database of urban planning. As urban rules become more complex, the duo proposed to develop tools that make it easier for the popular discourse to regain power over planning rules. For that, they developed a deep analysis of the rules, exploring alternatives and possible solutions to be applied by the population. The project also works as a collaborative tutorial, where the community can exchange knowledge on how to 'break' the rules without going against them.

In the architectural studio 'MUF', participation becomes part of the elements being 'represented' in architecture and participation traces become a set of elements in the final products' aesthetics. Moreover, according to the multi-disciplinary collective Assemble, participation not only enters the process as input, but it also becomes an output. According to an interview (Jones 2015), the aesthetics of products keeps a certain openness, as if users could see the way the object was produced and think 'I could do that myself'. Thus, the product becomes an empowering object, rather than a form to mask processes with appearances. Furthermore, based on the Assemble's Granby Four Streets project, the design process became the means of bringing a community together to actively produce their own spatial conditions.

In another initiative, Assemble transformed an abandoned petrol station into an ephemeral hand-built cinema. Developed in the do-it-yourself spirit, the project was built with ready-made resources, exploring the poetics of raw material. Successful architecture is revealed to be more an instrument of conviviality and encounter promotion than advanced technology and industrialized processes. Built using a volunteer workforce, the construction itself became an instrument to transform the place into a landmark for the community.

The collective Atelier Urban Nomads have created several initiatives that depart from engagement with local communities to transform and re-appropriate abandoned urban spaces. Using hand-made prototypes and small pieces of furniture, they could transform the perception and use of everyday spaces. In the initiative 'Jogos de Rua' (Street Games), the group approached different forms of recreational activities to explore new possibilities of emptied neighbourhood spaces. Through a series of workshops, residents were invited to participate in all steps of the process from conception to construction and use. Once set into practice, the process developed a new bond between the community and neighbours, ultimately creating new 'rituals' and meanings for places.

These experiences have inspired research and experiments for building a critical approach to the production of urban spaces, focusing on collaboration, design as a process and a subjectivity that recreates the values of the current fantasies of the field. Rather than viewing architecture as a static product, these approaches emphasize process, engagement, and the reappropriation of space as central elements of design. By positioning architecture as a tool for collective agency, they challenge the dominant frameworks that govern urban development, revealing alternative ways of shaping cities beyond institutional control. In this context, the transformation of public space is not merely a question of aesthetics or function but a site of negotiation, conflict, and potential emancipation.

The experiments and discussions that follow examines the potentialities and limits of these critical approaches, exploring architecture not as a fixed discipline but as an evolving field capable of responding to and shaping the social, political, and economic struggles embedded in the urban landscape.

Laboratory Life and Live Laboratories

The case studies explore design studios lived experiences in Brazil, UK, Belgium and Italy. This investigation was inspired by Latour and Woolgar's (1986) book *Laboratory Life: The Construction of Scientific Facts*, who approached scientists' practices from the perspective of an anthropologist who arrived in a remote tribe and tried to understand how social representations and relations interacted within the production of facts. In our case, it was done to develop a similar approach and distancing to help better understanding the theoretical device adopted by people to reflect on their own practices.

Therefore, the next sections will briefly explore a way of opening architectural studios' 'black boxes' to analyse the disciplinary aspects manifested in them, the dialectical dynamics likely to be explored and, finally, to explore how these aspects may suggest an ongoing paradigm shift.

Discipline will be herein explored based on the way distinction gives voice to some individuals and to others gives a duty to obey, as well as on how it produces specific subjectivities and results in a hidden creative surplus-value estrangement process.

Dialectics will be explored through action research based on flipped classroom, collaboration techniques and the concepts of *hacking* and *micro-utopias*. It will be done to explore the way it might create different ways of seeing things by revealing hidden properties of both society and space, and by imagining future possibilities and articulating new shared social realities.

Finally, the current study aims to explore how we live within conflicting paradigms such as invention and labour, competition and collaboration, distinction and collective dialogue, as well as seductive narratives and negotiated practices, thus, pointing towards the need of finding new collaborative approaches to the creative processes of transforming cities.

The Field Investigation

Field investigation comprised a series of 'constant comparative analysis', as proposed in the Grounded Theory approach (Guest et al. 2011; Glaser and Strauss 1967). Overall, it is a qualitative sociological method aimed at developing theories based on rigorous observations. To do so, four 1-week live-project workshops (two in London, one in Belgium, and one in Italy) and two 4-month studios in Brazil were approached through participant observation of their activities, which were recorded with photography and field journals. In addition, interview forms were applied to participants. All collected data were coded to enable retrieving key aspects and repetitions; secondly, they were thematized in concepts presented and discussed in conferences, before being finally reviewed in the present form.

Furthermore, it followed Tedlock's grounds for an 'ethno-sociology' capable of developing a self-reflexive ethnography (Tedlock 1991, pp. 78–80). It means that 'participant observation' became a more personal 'observation of participation' by keeping the crucial dilemma between 'participation' (which entails emotional involvement) and 'observation' (which requires detachment). Thus, its findings result from a unique and specific dialogue that dilutes the mediations shared by those 'who observe' and those 'who are observed'. According to Tedlock, this knowledge 'belongs neither to the realm of objectivity nor to that of subjectivity, but rather to "human intersubjectivity"'.

Every summer, UEL organizes a series of Live Projects called 'Construction Week'. In 1 or 2 weeks, groups of students from the master's and postgraduate architecture courses work collaboratively with staff and different communities to design and build a 1:1 architectural piece. These collaborative efforts aim to give students close contact with the concrete processes of producing architecture. The task, the context, the public, the learning opportunities and the demand vary, and the workshops range considerably in terms of scope. Nevertheless, these projects generally seek to engage students with communities and to extend the research of the university to the outside world.[1] This participant observation enabled the research to examine the ways in which the discipline operates and mobilizes its apparatus to reproduce urban interventions. As a result, rather than solving the paradoxes in the field, the hypotheses revealed further constraints of the approach.

6.2 Critical Observations and Stake Points

Several authors (including David Harvey and Whachsmuth 2011s and Boaventura Souza Santos 2007) have argued that we are experiencing a key moment in history, when great changes in social structure, culture, technology, and production, as well

[1] To avoid identification of personal sensitive data, the participants mentioned here will be anonymized.

as the threat of an eminent environmental collapse, force us to imagine new social alternatives. If we manage to open the black boxes of architecture, we could recognize how the architectural discipline reproduces social relations, both internally and externally (framing the possible life of ordinary people). Furthermore, if we open those black boxes to scrutiny, it could also become a way of envisioning new futures.

That would mean that architecture could not only design new things, but it also would have the potential to explore new ways of seeing things by: revealing hidden properties of social reality through new imaginative tools of representation and critical narration; by imagining other possible future scenarios based on using its design-thinking methodologies; and, finally, by refunding social realities in itself based on articulating virtual seeds and on building shared knowledge. Thus, we explored these issues in action research experiments, just as a chemist would mix different compounds in a lab, we mixed working hypotheses in the studio practice. The working hypotheses used were 'hacking' and 'micro-utopias' in interventions in urban spaces.

Furthermore, we adopted 'flipped classroom' methodologies, according to which hierarchy and knowledge production in the classroom is inverted, so the studio became a place of shared knowledge production. In addition, collaboration techniques, such as workshops, kanban and assemblies, were explored to emphasize the collective construction of the debates. Finally, the studios aimed at developing 'live projects' to overcome the traditional 'simulation studio'—where inputs are imagined and assumed—by focusing proposals on the strategies on the actual practice they resulted in.

Hacking Experiments in Urban Ecology

The industrial revolution confronted the social imagination with a fast-mechanical change in society; therefore, *history* was a concept vastly used to explain social transformations, through a linear universal concept of time (Lefebvre 1971b, pp. 17, 229–230). Recently, a sense that 'nothing is really changing' has subjugated many scholars with a feeling that 'the future is dead' (Wolfe 2012; Berardi 2011; Bauman 2010; inter alia). In an informational society, which is changing through reprogramming, how could social space transformation be conceived?

For Antonio Negri, we live in a cognitive labour age, associated with the automation of production, where revolution takes a new path. Therefore, he proposes a course of struggles towards reformatting and reprogramming platforms (of production, finance, logistics, and consumption) as a strategy of power subversion (Negri 2014). Nevertheless, Mckenzie Wark argues that despite all the changes that have taken place in the recent social space, the structural contradiction in capitalism remain largely the same (Wark 2004).

On the one hand, a social class has control over the means of production. On the other hand, there is the class that works and produces and that has become more engaged in abstract labour, after automation of production. But its creativeness is extracted by the soft machinery (protocol) of the ruling class (especially by copyright strategies). She names 'hackers' this contemporary class of dispossessed. In this way, Wark brings the idea of hacking to an everyday basis, i.e. creates a new class conscience (Wark 2013). It is in this sense that we aimed to explore the possibility of social transformation as a hacking experiment. But, how could hacking be defined?

Although many attempts have been made, instead of providing an ideal definition of hacking, we tackled hacking as an aesthetic procedure, as a way of 'seeing' differently and manipulating the system (of social space). In this sense, we started our experiments with a new understanding of the means of production in architecture rather than an objective aim. Rather than proposing structures, design was concerned with subverting existing public spaces rules to test the limits and paradoxes that could emerge from this experiment. Therefore, hacking in architecture was conceived as grappling with the pragmatic aesthetics of spatial frameworks.

Hacking aesthetics is structured into two key dimensions: attitudes and techniques, each with three aspects. The first attitude, the *Prank*, employs sarcasm to expose hidden truths through sharp critique (Kushner 2012; The Antics Roadshow 2011). The *Hammer* attitude embraces bold interventions, as seen in how Cory Arcangel hacked *Super Mario* cartridge (a brute-force intervention, 'messing around' the code, breaking things, through trial and error, until he got a punk-art effect in the game). The third attitude is *Collaboration*, where anonymous forums foster collective action. Among techniques, *Ready-made solutions* repurpose existing systems, akin to Christopher Alexander's pattern language (Alexander 1977). *Computer viruses*—including worms and Trojans—serve as tools for infiltration (Galloway and Thacker 2007). Finally, *Deep-in/interfacing* involves both decoding system vulnerabilities and designing interfaces for manipulation, with diagrams playing a crucial role. This evolving hacking aesthetic demands kludge wisdom and adaptive creativity, paralleling Ulysses' *mêtis* (Williams 2013b). Arguably, as architecture faces an unpredictable future, it must integrate new tactics to navigate emerging challenges.

Results in Hacking Privatized Spaces

The urban ecology module of the school of Architecture, Computing and Engineering (ACE) at UEL was coordinated by Roland Karthaus, Alan Chandler, and Anna Minton in 2013/2014. In this module, students developed theoretical research and a design experiment. The ecology of the urban space was understood as a complex set of relations that surround life in contemporary cities, amplifying the 'under-theorized' notion of the environment as only the 'physical'. In this sense, the rules, hierarchies, forms of power, modes of production, and even physical objects were understood as part of a larger (ecological) system (Karthaus et al. 2013). As a methodology, the module proposed a three-stage exercise: first, theoretical research; second, a 'performance' intended to test the operative concreteness of those systemic elements of the city; and third, the production of a material 'prototype' intended to interfere in those immaterial instances of the city to 'identify a subversive potential that could be exploited through a material construction' (Karthaus et al. 2013).

These objects were conceived as prototypes and models 'built to test a concept or a process or to act as a thing to be replicated or learned from'. The prototypes were to particularly focus on 'hacking' in the privatized urban space in regeneration areas of London, which was first explored via spatial 'performances'. Those performances were events produced by dissonant uses of the space. For instance, G.S. investigated the corporate control of images in privatized public spaces (Fig. 6.1). These spaces are deeply influenced by symbolic capital and are based on 'spectacle' strategies. Therefore, any photograph is forbidden, unless you have express authorization. In this sense, G.S. subverted the rules by simply drawing pictures in the space instead of using a camera.

Z.Y. (2014) worked with F.C., investigating a new privatized public space in Granary Square. As the space provides a series of measures to maintain a consumer-driven experience, security is orientated towards eliminating the practice of begging in the square. Nevertheless, the space is supposedly conceived as tool to enhance the 'playful' and 'free' interaction of its users. Exploring this limitation, Z.Y. dressed up in simple clothes and started to wander around the square asking passers-by if they were hungry and if they would accept a candy bar as a gift (Fig. 6.2).

After a few minutes, security arrived and asked him to leave the square, as his behaviour was prohibited. The student explained that he was not begging for food, rather he was giving. A superior was called; the activity was understood to be against the 'rules', and Z.Y. was warned to step outside the lines on the ground marking the division of public and private space or the police would be called.

These experiments helped reveal aspects of the contemporary city and served as a starting point to isolate and manipulate variables in these spaces; as a collateral effect, they also helped trace the first paradoxes and limits of architecture as a subversive tool. The activity, although in harmony with the intentions of the place, was somehow interfering with the expected uses of the space and with the intended appearance of its users. In this sense, the appearance of the activity was more important for the functioning of the ecology of the place than its actual agenda.

In the next phase of the research, Z.Y. investigated the diverse elements in the urban space that function as guides and instruments of control. As with the marks in the pavement called to Z.Y.'s attention, bollards are also important silent devices. Beyond the function of directing the flux of pedestrians and avoiding the flux of cars, they also create sensations of borders and delimit transitions.

As a type of prank intended to create awareness and shock, he proposed a prototype of a ready-made basketball net to be attached to these devices (Fig. 6.3). As they would stay within the limits of the public and the private, the objects would be outside the jurisdiction of the private, despite acting as virus that would infect the uses and image of the space and amplifying the 'playful' experience of the space beyond the planned limits. In contrast to 'security by design', the prototype was called 'fun by design'. By playing with the limits of the public and the private, he was able to insert a challenging experience into a controlled space.

Fig. 6.1 Urban Ecology 1. (Source: the author (based on information from Soiza 2012))

Stake Point

Experiments such as the one described above not only have pedagogical value in terms of confronting students with the relationship between the production of architecture and the reproduction of social relations but also have the potential to investigate how architectural skills and knowledge can be mobilized in practices that might subvert the smooth systems of social control. By investigating the frontiers of

Fig. 6.2 Urban Ecology 2.1. (Source: the author (based on information from Yoong 2014))

Fig. 6.3 Urban Ecology 2.2. (Source: the author (based on information from Yoong 2014))

the social rules engraved in urban spaces, design is mobilized as a means of change rather than a perpetual reinforcement of the establishment.

The 'hacking aesthetics' analogy is inspiring and helpful in unveiling abstract spatial codes, thus functioning as a means of entering the 'ecological system' of the city, which keeps subjects unaware of the built intentions and manipulations. Nevertheless, it might be too literal a conversion of the concept from the technological realm to the social. The urban space has too many particularities, and the dualistic distinctions and absolute categories of the digital realm—with abyssal choices between 0 and 1 or between hardware and software—are unable to tackle the dialectical dynamics of social spaces.

Furthermore, the belief that technology will 'redeem us from the old social structures' is an idea as old as the positivism of Comte (n.d. [1848]). This notion supports a blind belief in the capacity of technological advancements to potentially solve social problems. This assumption is shared by conservatives and some lines of critical thought that are now organized under the label of 'accelerationism', as if the solution to all problems can be found in the acceleration of progress (see Mackay and Avanessian 2014; Srnicek 2013; Williams 2013a; and, for a critical account see Negri 2014). Another point at stake in those experiments is the extent to which they become a form of 'protest' rather than delivering a transformation in the production of social relations. The power produced by 'protest' comes from its highly symbolic and representative dimension. As discussed in Chap. 1, social space cannot be equated with representation without falling into a deadlock.

Finally, the critical stake point relies on the distinction between the 'tactical' and 'strategic' agencies provided by these experiments. As in a game of chess, the strategies are the general picture concerned with the structure of the game, and tactics are the step-by-step actions. In this sense, hacking can be understood more as a tactical instrument and less as a 'whole picture strategy'. These series of critical points lead our research to the next investigation. Can architecture subvert the traditional duality of scales (the big and the small, the local and the global, the part and the whole) and produce interventions that, although small in scale, have a concrete impact on the whole?

Micro-utopias on the Street

Cultural expectations, scientific developments, technological instruments and the new social conditions might be changing the architectural field of work. Architectural theory has traditionally defended hierarchical forms of work, transforming architecture into an instrument for maintaining social inequality. In parallel, critical theories of architecture theory have either fought against architecture itself or searched for alternatives outside architecture. Nonetheless, could architects emancipate themselves and recreate their set of skills to develop emancipated and autonomous practices?

Although architects have not been able to change society, the conception of their role may have contributed to this failure. Since the Renaissance, the idea of rationality has played a strong role in the hopes of humanist realization. Since the Enlightenment, the idea of technology has also come into play, resulting in a strong faith in progress. If the ancient regime, the court (which included architects) held divine rights, the French Revolution declared that all human beings are born equal. Although this is one of the greatest inventions of all times, the second right surprisingly declared as 'natural' the right of property, along with liberty, security, and non-oppression. This caused a great conflict of interests. For more than 200 years, Western societies have not been able to set into practice the claims of equality, fraternity, and liberty. As a mirror reflecting these contradictions, architecture's own utopias have faced difficulties in becoming realized. Nevertheless, a new social role for architecture is emerging, and many debates call into question what should be destroyed, what should be built and, furthermore, how it should be done.

Social emancipation has evolved alongside shifts in production. While mechanization and automation promised liberation from labour, they have instead intensified mental work due to production's growing complexity and instability (Crary 2013). This fragmentation isolates workers, reshaping class identities and enabling new forms of control and hierarchy. Zizek (2008) highlights a new politics of violence, where media perpetuates symbolic terror, while States hold monopoly of violence. Anna Minton (2012) argues that fear facilitates heightened social control, evident in widespread surveillance, and security protocols. Meanwhile, Wark (2004) asserts that copyright laws reinforce hierarchical control over intellectual production, maintaining social stratification. Despite these mechanisms, mass uprisings have challenged the status quo. The 2008 financial crisis ignited the Occupy movement, followed by the Arab Spring (in 2010) and Spain's Indignados (in 2011–2012), fostering renewed debates on social transformation, although all the challenges are yet to be overcome.

On the one hand, these 'masses' were relatively ineffective in reclaiming new social gains within the old European democracies. Democracy is facing not only a crisis of representation, but also the emergence of deep-fake-news, neo-fascism, and criminalization of dissent. This situation has created arguments for new forms of democracy and social control. On the other hand, new forms of social activism are being born with the creative use of technologies of communication. This process has boosted new forms of identity and new forms of social articulation and struggle.

Urban interventions and redevelopments increasingly blur the line between architecture and broader urban policy. New technologies reshape urban space, dissolving traditional boundaries between city and world. Lefebvre (*Architecture of Enjoyment*) argues that architecture can manipulate social rules at the concrete scale while influencing the broader urban fabric. This challenges traditional hierarchies and urban governance. The concept of *micro-utopian* architecture emerges, where local interventions become tools for broader social transformation, bridging individual spatial practices with systemic change.

The idea of a micro-utopia was approached through two main aspects. On the one hand, it addresses a fractal and kaleidoscopic conception of action (see Soja 2000). This idea builds upon the concept of micro-politics, stating that power is assured by concrete strategies in everyday life and the concrete body, both physical and psychological. On the other hand, this idea also challenges the traditional sense of scale, which opposes the macro and the micro, where the latter is only a reflection of the former. If the scale of the 'close' is also the scale of the concrete realization, punctual interventions might realize changes in the 'kaleidoscopic image' of the whole. In this sense, architectonical interventions could develop autonomous social relations on a concrete scale, as occurs within institutions (see Lefebvre 2014), providing the grounds for new forms of social struggles.

Mannheim would argue that ideology is the unreal, rather than utopia, so Mannheim inverts the common association of the pairs 'ideology-utopia' 'possible-impossible' (Mannheim 2000, pp. 173, 175). For him, ideology (i.e. the hegemonic cultural logic) presents itself as a reality 'de facto' although it has never been realized (acting only as a form of legitimating the existing order), whereas utopia is an idea that envisions a real possibility of rupturing with existing order. Thus, utopias should be understood as the construction of real possibilities departing from the existing context, which is deemed to be changed. In this sense, Lefebvre (2000) suggests a method of 'transduction'—the production of a virtual possibility—with which architecture could act in the production of possibilities that escapes the realized possibilities.

Micro-utopia is framed as the process of reclaiming the commons. Galloway (2004) describes *temporary autonomous zones* (TAZ) as counter-spaces of concrete freedom, where resistance emerges and dissolves before the State can suppress it. Hackers, he argues, operate within these spaces, embodying a form of ephemeral insurgency. Sande (2013) expands on this by defining micro-utopias as *acting as if one is already free*, drawing from Foucault's micro-politics. This aligns with Nietzsche's (1987) concept of *active nihilism*, which rejects ultimate values in favour of self-invented ones implemented in reality. Thus, micro-utopias are both a struggle for freedom and an act of creating freedom, functioning as experimental spaces of social and spatial transformation. To evaluate architecture's role in micro-utopian projects, we analysed workshops creating small prototypes as urban *TAZ*.

Results in the Street

Wood Street, a historic shopping street in Walthamstow, London, is flanked by a council estate and a redesigned twentieth-century plaza. The local council's *Area Action Plan* (Borough Council of Waltham Forest 2013) guides its 15-year regeneration. Between 2009 and 2011, consultations included meetings, retailer surveys, and an *Urban Design Framework* featuring a model highlighting intervention sites. Residents, stakeholders, and Ward Members attended exhibitions to review proposals. The plan outlines 6 clusters and 19 sites, including *Cluster 4*, which contains *Site 10—Marlowe Road* and *Site 11—The Plaza*. It proposes demolishing the estate, retaining Northwood Tower, and rebuilding a mixed-tenure development. Additionally, *The Plaza* is set for redevelopment, integrating flats above shops.

The consultation process was deemed insufficient by the organized community, as it primarily legitimized pre-existing regeneration schemes, similar to those in previous case studies. The methodology created an illusion of consensus by aligning with planning requirements while reinforcing predetermined solutions. The official report noted residents' support for local shops, independent businesses, and improved public spaces. However, the project's conclusions ultimately led to the demolition of social housing and the privatization of public spaces. While the report presented appealing visuals, its reasoning appeared deceptive, failing to align participatory feedback with actual outcomes. In response, *Architects E17* (a non-profit) and *Wood St. First* (a community group) initiated an independent engagement process to engage the community:

> AE17 were keen that local people should be positively engaged in the future of the area and wanted to demonstrate how this could be facilitated. With a small grant from Wood Street First and the RIBA, AE17 made a post box, called Dolly the Trolley who spent the summer in various locations and events around Wood Street collecting people's views on the area and ideas for the square. Workshops were held with various local groups, including the Soul

> Project and Frederick Bremer school, the local secondary school located north of Wood Street off Fulbourne Road. On Saturday, 6th September, an ideas day was held where local people and AE17 together made models of temporary interventions that could create new possibilities for the use of the square. Representatives from Waltham Forest Council and Wood Street First acted as a jury to select the best ideas (Karthaus 2014).

This mobilization attracted the attention of the local council, leading to a partnership with UEL's School of Architecture for a Live Project. Its mission was to design and build temporary installations in Wood Street Plaza that reflected community aspirations and expanded its use. The intervention would develop gradually, culminating in a final Saturday event. The installation's duration remained uncertain, depending on local adoption or removal (Karthaus 2014). Seven students, two architectural researchers, one social work researcher, and a tutor participated. Local suppliers provided recycled materials, RIBA contributed community project funding, and a local school offered workshop access. The students launched the design process at a café and later experimented with cardboard box performances, engaging locals and testing how small interventions could influence the space's activities (Fig. 6.4).

Fig. 6.4 Wood Street experiment 1. (Source: the author)

The students began to explore the material that they could use and design possibilities. Five days later, they presented some ideas to a local jury at a community centre and brought a wood frame to the plaza to support boxes that would hold proposals from locals and display historical information about the site (Fig. 6.5). The handling of different interests remained concealed, and the overall discussion centred on empowering the community in the discussion.

Students from the local school helped construct pieces of the design and participated in the assembly testing of the structure. Leaflets were distributed for the event, and the local café provided food. Parts were produced during the week, and the structure was raised in the morning of the event. A small group of people gathered around the structure, and some curiosity was generated for the event, which kindled some heated debates (Fig. 6.6). On this occasion, the circumstances of the Area Action Plan and the future of the plaza were discussed. Proposals were collected from the local community, and kids were encouraged to contribute to the project with their drawings.

At the end of the event, when the structure began being disassembled, a group of kids furiously demolished all the boxes. Something was in the air, exposing the paradoxes brought to light through this experience. As the structure was completely removed, the plaza remained just as before, and one of the participants asked: 'Did all this work change the square in any way?'

Fig. 6.5 Wood Street experiment 2. (Source: the author)

Stake Point

As part of the mobilization led by Architects E17, this process continued in the form of other initiatives. Especially interesting was the fact that this project lead to an initiative that introduced a web map called 'Where in Wally?', where locals could share places that they considered important to the community. According to the Localism Act of 2012, these places can be listed as 'Assets of Community Value' and thus (hypothetically) be protected from market forces. The ephemeral 'architectonic' device built in the square was more a means of mobilization than an actual product—more of a process than a thing.

Could an architectural procedure have changed this space by not actually changing it? In the case of Wood Street, architecture emerged as a means to raise awareness about the regeneration process underway. It became a device that temporarily shifted the actual centre of decisions, bringing into action a collective subject. In this way, the fact that the plaza remained unchanged—not regenerated—was, for a time, the direct outcome of the design process itself. Thus, the potential fate of Wood Street was suspended, and this 'architectural' intervention existed in a virtual sense that temporarily had real consequences (Fig. 6.7). However, the regeneration was only delayed for 4 years. Without the sustained momentum of local resistance,

Fig. 6.6 Wood Street experiment 3. (Source: the author)

Fig. 6.7 Wood Street experiment: the unchanged square after intervention. (Source: the author)

the collective capacity that once halted the project gradually dissolved. In the end, the movement lost the battle, and the regeneration proceeded. Yet, this episode demonstrates that architecture can become a political tool by creating space for dissent and collective agency.

This experiment points to a core stake point. The workshop did not provide 'architecture' as a 'thing', instead, this workshop product was an event in the sense of Alain Badiou (2012). For Badiou, an event is the transformation of 'appearances' by the emergence of a collective subject, where the reality of things is suddenly changed in a 'moment of truth', when a collective joins together in a singular political process. 'Things' become a process. To understand how an empty square might be understood as architecture, we would need to radically transform the idea of 'architecture' from a structure to an operation.

Live Projects and its Contradictions

Specifically, in academia, a form of architectural pedagogy has emerged that involves concrete projects in what have been called 'Live Projects'. Working along the margins of education and building practices, these projects engaged education in communities and local interventions (see Harriss and Widder 2014).

Brown (2014), exploring the classical tension between 'product' and 'process' in architectural education, argued that Live Projects provide a base from which to surpass the classical pedagogy of architecture based in the design studio. For him, Live Projects allow to break with the binary teacher/student and to avoid the conception of a 'banking education', where knowledge is deposited in a passive student. Such projects point towards a critical pedagogy that not only applies an experimental learning model (experience > reflection > abstraction > experiment) but also engages architecture as a collective process of production. Therefore, students can learn while questioning the existing 'roles' and 'positions' of the architect in a concrete 'participation in the world'.

Gloster (2014, pp. 45–55) finds this 'guerrilla practice' helpful in supporting the construction of the students' skills and profile, even as defined by the Royal Institute of British Architects (RIBA) accreditation. Additionally, Chandler (2014) notes that the aims stated by RIBA criteria of professional standards are more centred on 'reflection' as separated from 'doing', so he argues for a 'building of reality' and the 'reality of building' as a form of breaking the long-standing disengagement of the field. This practice of Live Projects has found fertile ground at UEL, as this institution has a great interest and a leading position in social engagement. For example, the design collective Assemble (leaders of some of those Live Projects) won the most prestigious British prize for innovative artistic practices (the Turner Prize), showing how these practices are on the cutting edge—not only in architecture but also in other disciplines.

As a counterexample, the workshop 'Into the Green' held in London also implemented an intervention in a public space during an event that mobilized the local community. Consequently, an office was opened at the university to develop interventions in that space, so that the ephemeral community activities held in it could become permanent. However, the office was dismantled due to the university's bureaucracy, the community became disengaged, and it put an end to the process.

Another workshop held in Belgium shows more blurring of these lines. The workshop aimed to identify local desires and elements that could be realized, but students struggled to determine the limits of their agency and were continually concerned with meeting their teachers' expectations. This experience brought further questions to light. It was paradoxically unresolved to what extent the resulting objects were more than sculptures representing specific community interests and to what extent they were merely educational activities. Additionally, because the objects symbolized community participation, they were no less constrained by the prevailing ideology than the average 'good citizen.' This raised concerns about whether architecture was effecting change or merely reinforcing the *status quo*, ultimately questioning whether it had become an empty community game.

Departing from these experiences, several studios in Brazil worked with the idea of micro-utopias by taking into consideration immanent potentialities and by exploring their development with the aim to break up established ways of seeing social structures. Nonetheless, besides the utopian intent, it was quite common for students to rather develop dystopias, a fact that can be read as sign of our times: it is easier imagining society's collapse than any alternatives to current scenarios. Accordingly, our social-environmental issue is also an aesthetical issue: how can we collectively see the solution for unsustainable development? Cities' aesthetics goes far beyond simple 'beautification' since it establishes a way of perceiving and establishing relevant issues for a given community. Thus, urban space creates an 'aesthetic field', i.e. a setting that expands, or limits, the vision shared by a group of citizens.

Another studio in Goiânia (Brazil) worked in partnership with the local environmental agency to investigate and propose solutions to help recovering its main river. As the analysis went by, the collective conclusion was that it was not possible to develop one single project to solve issues faced by the river since they involved all tributaries of the river and the whole city's relationship with them. Therefore, the proposal lay on developing a series of actions to gain public attention to these issues; thus, architecture was no longer a means of designing things, but of making visible the hidden city-local nature relationship. In addition, a collaborative web platform and an NGO were launched as a means to permanently address and share proposals and visions to build new paths.

Stake Point

Finally, a survey applied a form to collect the opinion of students about the experience in Wood Street. The 'labour condition' issue was assessed as the best dimension of their experience, with 80% considering it positive. This topic assessment involved the role played by participants in the production and sharing of ideas, the openness of debates, and the democratic decision-making in comparison to previous academic experiences.

Surprisingly, students did not consider this experience significantly different from other past practices. This finding suggests that the actual work experience is always collective in practice. Although the narrative in the field is that the production is done by the main architect in the office, in practice the involvement of trainees in the design process resembles the collaborative methodologies explored in the workshops. This might be the case of any office, where the collective work is anonymous, and the main architect is the reference—again, Assemble functions on the opposite pole, where the work is done in a collective process, according to which, all activities from design process to everyday reproduction are shared.

Form results for the 'delivered products' dimension were mostly paradoxical given the Live Projects' intentions. Based on students' perception, products had weak positive impact on the place (only 19% above neutrality), the connection between creativity and construction was limited (19.8% above neutrality), and the product was quite similar to that of a conventional architectural practice (3.4% above neutrality). However, the analysed interventions were considered a 'valuable project' (31% above neutrality).

This condition points out that the architectural studio has a long tradition and that even counter-practices remain entangled in this long tradition. Thus, it is not the case of advocating for fully free practice, but, as Hanna Arendt (2006) suggested in a different context, we should focus on the continuous fight for freedom, on the continuous deconstruction and critical awareness of limits imposed by the discipline, and on potentialities enabled by a dialectical attitude towards design and social structuring.

Furthermore, in one specific workshop experience—which we will not identify for privacy reasons—we noticed how distinction and subjectification remained among several subjects involved. Distinction still plays a key role in placing agents in different social positions, so both teachers and students' voices are structured in a hierarchical manner among themselves. According to Bourdieu (1996b), pressures mediating the production of different works are both internal and external to the associated field. Moreover, they relate to the symbolic capital associated with participants' previous creative experience. In addition, the routine imposed by institutions on agents establishes a symbolic order, a valuation circle, a publicity level and a definition of legitimacy. Distinction is the first step in the subjectification process since it establishes who has a voice. Precisely because other subjects crave for the same distinction, a continuous circular reinforcement is at play.

That particular case study helped investigating how architecture is not just a practice, but also a subjectification process based on narratives capable of reproducing estrangement before requiring submission and providing mastery. Thus, these narratives stand on a previous acceptance of instrumentalization.

Notably, instrumentalization resulted in a continuous authorship appropriating process that can be called 'surplus of creation' estrangement. The dominant group not only proposed ideas to be followed by students, but also appropriated students' work. During a design exercise, a door was to be created by subtracting modules from the pavilion's wall. The tutor insisted she had already solved the problem and that students only needed to 'draw it'. However, practical issues emerged—her symmetrical sketches were unfeasible, and the proposed door position made the cantilever unstable. Despite this, she dismissed student concerns with irony. Aware of being observed, students remained agreeable but later worked independently to find a viable solution by slightly rotating the entire structure. Once resolved, the tutor claimed they had finally understood her idea, despite no prior solution existing.

In another case, the tutor repeatedly changed the ground floor plan designed by students although his changes made no practical difference. Better solutions proposed by students were re-sketched until the tutor felt to own the solutions. Again, in a paradigmatic case, a design was structured with larger modules in the base and with smaller modules at the top. The tutor took the model and inverted it, asking a fellow teacher, 'Should we do it like this? It will create an optical illusion that the structure is bigger'. Although the intended optical illusion would technically be delivered by the original proposal, students (tired of continuous conflicts) accepted the tutor's suggestion as the 'stroke of a genius'. Their initial design was magically reworked by the genius (a change without changes), and the whole thing now belonged to his act of 'creation'.

Thus, the working subjects became instruments through a double step: first, they became estranged from their own background to fit the distinction structure in the classroom; later on, they were estranged from the product of their own work. This process not just fulfilled the desires of the group in power, but also genuinely created new solutions, which were later assumed to result from their obedience. Thus, the disciplinary black box of architecture not only produces distinctions, but the way subjects see the world, their place in it, and how they can act in it. Those are the cogs and gears of this abstract machinery.

6.3 Exploring Contradictions in Counter-Hegemonic Practices

Finally, to move forward in the analysis of architecture and the reproduction of social relations, it will be important to critique and overcome the traditional forms of historical narrative and valorization of architecture (unveil its collective subject) so we can investigate how emergent and dissentious practices could challenge different aspects of the reproduction of discipline.

Architecture as a Process

The Wood Street fieldwork added a conception of architecture in a completely new dimension. When the housing block and a local square were facing the risk of demolition and privatization, a mediation between different groups (University of East London, E17 Architects, Wood St. First community group, WF Council, and the RIBA) enabled the production of an ephemeral architectonic structure built in the square. This structure functioned only as a means to mobilize a series of activities and to enable the participation of the community in the debate about the transformations to come in the local area. The object produced was less important than the event it created.

One could argue that after the event, the square remained physically exactly as it had been before. In classical architectural ideology, this could not be understood as architecture, as it did not produce a permanent structure. Only with the understanding of architecture as operation, as ontogenesis, could one understand it as an operation that enabled further operations. This mobilization led not only to the interruption of the plans to regenerate the square but also to other initiatives to protect the important spaces for the community. If architecture can be understood as the creation of counter potentialities, then it can change reality by opening new horizons: a potential, yet real, existence. In that case, the architecture of the city is neither codes of beauty, nor a style, nor an affiliation with a school; it is a political praxis.

After all, and at least, these case studies enlighten architectural production as process, in contrast to narratives about free individuals acting with free autonomy to develop new architectonic things is the process of continuous transformation, where matter is not passive but holds multiple potentials for change. It interacts dynamically with external forces, undergoing structuring processes that shape its form and function. Things come into being through tensions between constraints and possibilities, where external forces shape and react against the internal tendencies of a system. Unlike fixed entities, these living systems exist in an unstable state, continuously adapting and evolving. Rather than being finite, existence is limited, engaging with external elements to reorder and transform itself. This relational process enables genuine creation, invention, and the emergence of new forms.

Thus, looking back to the current object of study, it is possible to see that architecture is not a substance. It is not an immutable essence, something based on some specific features regardless of its actual production process. Moreover, architecture is not a set of formal properties abstractly defined based on a set of ideal features. Rather, the current study explored how the discipline of architecture limits its being and how the dynamic process of dialectical transformations works.

All Good Design Is Collaborative Design

The Competition (2013) is a documentary movie focused on telling the story of several star architects competing for a project in Luxembourg. Among several similar scenes, one shows the moment when a team of architects explains their proposal to the main architects, and one of them takes the model, turns it upside down and asks whether that was not the best solution. Thus, several hours of hard work by the team were 'magically' estranged from the collective effort to become the genius stroke of one single man (a repetition of what we saw in practice).

Another key argument that can be built in these case studies is that architectural production is a long process, just as the previously used brick example, since it involves traditional knowledge and several hours of creative and design-thinking experimentations. Nonetheless, both narratives and field structure set our perception to see individuals as creative drivers. Thus, what is essentially collaborative is captured in an abstract machinery that is reproduced by the discipline, and it ends up being perceived as a single-person invention.

Although this machinery operates in quite abstract and subtle terms, the aim of the current study was to investigate how distinction establishes a hierarchical structure in creation processes, how subjectification establishes a thought pattern by placing different subjects in different positions where they can act in and by limiting how they think and act in the world. Finally, this machinery produces the estrangement of collective products, in such a way that subjects involved do not see the product of their work as their own.

This is only possible because field narratives depict architects as individuals in competition, thus enclosing their subjectivities. That hides the actual collective production of architecture. If hegemonic narratives create this process of estrangement, we need new counter-narratives to oppose them. If fantasies of individual creation seed competition between architects, we need a new paradigm for envisioning the collaborative dimension at play. If we should care more about architects being explored in unpaid extra-hours, we need new perspectives to envision power relations inside the discipline and a new consciousness of architects as labourers. Although the collective production is captured by this abstract machinery, contradictory forces are at place. In this sense, there is space for a dialectical paradigm to be built to rethink our practice as fundamentally collaborative.

In this dialectical practice, architecture plays the social role of revealing, reimagining, and refunding social structuration. Through representation, diagrams and spatial analysis, architects can make aspects of reality that were not seen by common people before visible. By using design-thinking methodologies, architecture can radically reimagine the foundations of social space. And by setting these ideas into practice, we can collaborate to create more sustainable social structures.

Ultimately, this contradiction between disciplining forces and dialectical processes represents a transition time of conflicting paradigms. Between invention and labour, architects do not see themselves as workers; between competition and collaboration, architects see themselves as individuals in fierce fight; between distinction and collective dialogue, internal hierarchies give voice to few and estrange others; between seductive narratives of great masters and hard-time assemblies of negotiated practices, architects still tend to use the few symbolic power they still have.

On the Road to a New Subject of Spatial Production

A classical essay written by Judith Butler (1988) aimed at denaturalizing the understanding of our own body and sought to understand gender as performance, or as socially constructed phenomenology. Helene Shugart (2001) also explored how femininity in women and masculinity in men can be understood as social performance by de-fetishizing their mechanisms, based on parody studies. The aforementioned author sees the possibility of denaturalizing the nature of gender and, consequently, of reconstituting desire, by doing so. Arguably, we could also denaturalize architecture, and the creative labour involved in its production to set the ground for seeing architecture as it performs.

According to Bronwyn Davies (2006), who analysed Butler's relevance for education, subjectification is the process that simultaneously establishes mastery and submission. To master a given topic, students submit themselves to the perceived order of distinctions. It is herein possible tracing a link to Althusser's (1971) concept of 'interpellation' (an identity mirror). Interpellations create the subject because they create an image of the world and place the individual in a relationship with this world, and in a position based on which he/she can act in this world. Interpellations functions through ordinary objects address a subject setting ideological expectations and interaction rules, incorporating ideological discourses (such as the placement of chairs in a studio establishes institutional roles). Thus, these ordinary objects become embodied ideologies shaping internal subjectivities. Therefore, the material condition of a given production mode has a dialectical relationship with subjects' cognitive awareness of their place in this condition. Accordingly, discipline and distinctions produce subjectivity by framing the way architects see the world, their position in it, as well as how they understand the way they can act in the world.

Following Jason Read, the idea of the 'production of subjectivity' implies a double meaning: it is both something 'productive' and something 'produced'. Subjects emerge always in the context of a collective of subjects, i.e. there is no subject detached in an abstract empty space. This means that subjectivity is formed by elements a priori (language, culture, structure, social expectations, among others) 'externalised in machines and internalised in concepts, habits, and ways of thinking' (Read 2010). The dialectic operation of these abstract machinery of subjectivities creates a problem that is specifically political because they act across the whole society, seizing for individuals what was once formed by collective efforts.

Nonetheless, although the traditional architecture discipline encloses subjects in individual creativity and invention narratives, the observation of participation suggests that the most creative moments in architecture emerge from long processes of exchange, discussion, and maturing of alternatives. Thus, collective subjectivity is gradually produced, and it allows eventual individual creation to emerge from this transversal experience. Furthermore, according to Simondon, subjects can change, therefore, they are in a metastable condition that includes non-actualized potentialities. These are non-actualized potentialities that exist in all subjects, and are formed of internal tensions and multiplicities, besides being in constant exchange with their surroundings. It is through these tensions that the architectural discipline operates its abstract machinery, enabling the development of both a given architectonic object and the production of architectonic subjects, themselves.

Chapter 7
Subject to Change: Imagining Transitional Subjectivities

Abstract The last chapter dwells on the case studies reported in the previous chapters to develop an alternative way of understanding agency beyond the concept of individual subjects. To do so, it develops further Simondon's concept of ontogenesis (being as becoming) with political and ethical approaches. It changes the approach to the narratives about the historical development of the field and enables understanding architecture as abstract machinery, rather than as product of individual geniuses. This factor allows understanding collaborative and counter-hegemonic actions from a brand-new perspective.

Keywords Subjectivity · Agency · Ontogenesis · Individuation · Transindividuality · Metastability · Structuration · Reification · Cartography · Social transformation

7.1 Beyond the Individual

If Deleuze once asserted that 'Nothing is harder to define than the individual' (in Toscano 2006, p. 157), this is an old problem that has a long development in the history of ideas (see Simondon 2013, pp. 357–520). Here, we do not aim to retell that history, only to map the differences that might build an understanding of the relations between disciplinary processes and the reproduction of the subjectivity.

Being Becoming

> There are two views according to which the reality of being as individual can be approached: a substantialist view, considering being as consisting in its unity, giving to itself, founded upon itself, not generated, resistant to what is not itself; and a hylemorphistic view, considering the individual as generated by the encounter of form and matter. But in those two views there is something in common (...) Departing from the created individual, the effort is to reach back to the conditions of its existence (...) it is the individual as long as constituted individual the reality of interest, the reality to be explained (...) *Such a perspective of*

C. V. De Lima Amaral, *The Urban Reproduction of Subjectivities*,
SpringerBriefs in Geography, https://doi.org/10.1007/978-3-032-01465-8_7

> *research gives ontological privilege to the constituted individual*. Thus, it risks not approaching a truthful ontogenesis, of not positioning the individual inside the system of reality in which the individuation is produced (Simondon 2013, p. 23, *our translation*).

It is in this way that Simondon begins his thesis, approaching the history of the concept of the individual—displayed in full in 166 pages of its Appendices—in two groups. These two views together represent a model that presupposes an opposition between being and becoming, even if becoming might be conceived as an ability to reproduce its own reference (Simondon 2013, p. 25). For Simondon (2013, p. 49), to overcome substantialism, thought must depart from the process of the genesis of the individual, from the action of individuation (i.e., becoming being, without the 'and' in the middle).

If one considers becoming as a dimension of being, then relations between its interior and its exterior are part of being, without naming any new obscure substance (as in 'behind the mask of appearance'). The living individual is the one who continuously re-enacts the operation of its becoming, its individuation in a 'theatre of individuations' (Simondon 2013, p. 29). For Simondon, seeing reality as becoming transforms the 'finite' being of substantialism into a being that is instead 'limited' (Simondon 2013, p. 93). Being, as such, can be conceived in dynamic terms, as an undefined being bearing potential energy bigger than its factual actualisation. Thus, the idea of a 'limited' being can acknowledge how being can relate to outside matter and how it can incorporate, reorder and transform itself. For Simondon, only in this way can transformation (creation and invention) be consistently conceived. The notion of a finite and eternal being excludes the possibility of change, since what is considered 'fundamental' is taken to be inherently immutable. Instead, Simondon argues that limits are not fixed but part of a structuration process, forming a relational border—where exchanges occur—between inside and outside (Simondon 2013, p. 95).

Thus, Simondon changes the inquiry from an ontology of universal and eternal beings towards an 'ontogenesis'—the study of the creation of being, i.e. becoming. In this conception, for individuation to be possible, a preindividual field of possibilities that exist beforehand must be taken into consideration. In this sense, for Simondon, the individual is only a 'phase' (an 'actualisation') in a series of potential individuations. That idea explains how a subject can be formed of internal contradictions. Furthermore, the being as a structuration process for creating a limit, a border, keeps a continuous process of exchanges between an individuated individual, its externality and other individuals. This continuous 'theatre' of exchanges forms a realm of 'transindividuality'. Sometimes, non-actualized potentialities (contradictions) inside diverse individuals actualize together, forming transindividualities (i.e. collective subjects).

To overcome the hylemorphism of matter and form one must include a third term: energy (arguably, action, labour). Only by adding energy to this formula is it possible to think of being as a dynamic system, as a process of relations and interactions (i.e., the being as becoming), and even as the product of an intention, of a work, of an act of production.

Simondon elegantly illustrates this three-term 'ontogenesis' with a long consideration of the 'taking form' of a brick (Simondon 2013, pp. 39–58). The 'clay' as matter is not simply passive raw material; it has multiple possible transformations; it has aptitudes and tendencies. However, the clay is already a processed material, for which the grains were selected and to which the right amount of humidity was added. Additionally, it was collected and transported. Its properties and identification were only made possible by means of a long process of knowledge. As for the 'mould', it is not simply an abstract shape, conceived in advance by an intellectual process; it has a procedural role in imposing a limit on the transformation of the clay. This is an active action; more precisely, it is a reactive action of an equal and opposite force in relation to the one exercised by the clay on it.

The process of individuation of the brick is, so far, already composed of a dynamic system of interaction, with potentialities and forces interacting to produce a final state of stability. Nevertheless, one should add to this scenario the actual work of the artisan, which separates, discharges and presses the clay, while also using complex and subtle artifices to open and close the mould and give it perfect geometrical limits. Nevertheless, Simondon notes that the individuation of physical and technical objects occurs (usually) only once. Thus, such individuation is relatively stable and cannot transform itself. It can only degrade. For Simondon, classical philosophy knew only this type of stability, but the thermodynamic concept of 'metastability' offers new possibilities for understanding further individuation processes: the living individuations.

Simondon distinguishes among three 'phases' of being: the physical (as we saw in the case of the brick, entailing the formation of a stable limit); the biological (as in living beings, which are in a continuous process of individuation that is therefore called a 'theatre of individuation', where the internal is a kingdom inside a kingdom, in continuous exchanges); and finally the Psychic/collective individuation. This last phase of being explains the relation of individuals and society as a sort of double (perhaps multiple) individuation, whereas individuals individuate themselves as singular and collective individuals at the same time. This double individuation is what Simondon (2013, pp. 260–266) calls individualisation (though, as I will argue, we could call it subjectification).

With this torsion of ontology, individuality is the result of a preindividual, which is actualised transindividually. In the diagram below (Fig. 7.1), preindividuality is the dimension of being that comes before an individual is individuated. In this sense, each individual is a structured process of limiting. It bears inside itself contradictions and possible individuations (expressed as S_3 in the scheme). By means of an operation (O1), a mediation (M) and an interaction (I) are made possible between different orders of magnitude (also referred to as preindividual scales), where this process is called transduction (t). Thus, structure is only an instant, a means between the first operation and the following being becoming (O2).

Combes (2013, pp. 14–15) explains the change produced by this scheme—named Allagmatic, meaning a theory of operations—by introducing a theoretical analogy. She argues that Marx traced the constitution of capitalism by the substitution of the formula C-M-C for the formula M-C-M. Here, C stands for commodity

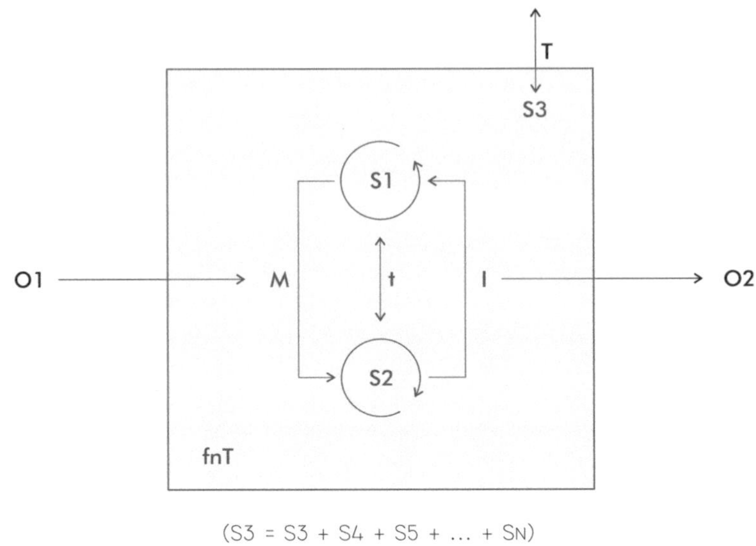

Fig. 7.1 Ontogenesis of individuations. (Source: the author (based on—but not equal to—Simondon 2013))

and M for money. Initially, in the former mode of exchange, a commodity was traded using money, which could later be exchanged for another commodity. Once money becomes the beginning and the end of the process, with the commodity only being its means, Capitalism is born and starts to reproduce itself. Similarly, she argues that Simondon changes the original notion of S-O-S (where S stands for structure and O for operation) to the notion of O-S-O. In this sense, 'operation' becomes the beginning and the end, and the structuration is only an instant when form and matter coincide.

Mixing thermodynamics with philosophy, Simondon (2013) conceives of 'being' as a process of information (*mise en forme*—to shape). This information is a provisional structuration of dynamic forces into a metastable resolution. Beyond the simple actualisation of inner virtualities (as a seed that becomes a three), his theory is a conception of individuation as a process of structuring disparate forces. Thus, the individuation is the creation of a mediation that articulates a field of conflicting energies (or disparate orders of magnitude). In this sense, the inquiry shifts from an investigation of the internal (individual) as opposed to the external (society) to an investigation of an energetic process in constant actualisation.

For a Copernican Revolution of the Individual

The final paragraph of Freud's (1957 [originally published in 1914], p. 101) analysis of the fundamental 'narcissism' of the unconscious starts with a remarkable conclusion: group psychology could be understood on the basis of individual narcissism as long as the collective is understood as a manifestation of homosexual libido, or ultimately, self-love. Following his text on Narcissism, Freud went on to write *Group Psychology* and *Society and its Discontents*. Jodi Dean (2016) offers a solid analysis of how Freud's 'sociology' reduces the collective (specifically the crowd) to a collection of individuals. As Dean asserts, all this deviation is important to understand how

> Not only is agency privileged over structure but the presumption that agents are individuals formats the alternative of autonomy or subjugation as an opposition between individual and collective. Collectivity comes to be associated with constraint, with preventing rather than enabling creativity and initiative. Liberal political theorists explicitly construe political agency as an individual capacity; others take the individuality of the subject of politics for granted. I argue that the problem of the subject is a problem of this persistent individual form, a form that encloses collective political subjectivity into the singular figure of the individual (Dean 2016, p. 364).

Dean (2016, pp. 368–369) asserts that this representation is itself a fantasy that places what one can only do among 'others' into an imaginary ego, thus making twin concepts of agency and the individual. Dean notes here how Freud's theory produces an 'enclosure of the individual' (Dean 2016, p. 371). Freud's analysis of the crowd, as examined by Dean (2016, p. 372), was deeply influenced by Le Bon's *The Crowd: A Study of the Popular Mind*. Le Bon, described as a 'pessimistic conservative' (Dean 2016, pp. 374–375), framed the crowd as a primitive force, threatening the social order. While Le Bon viewed the crowd as a process with an unconscious rooted in hereditary traits (p. 376), Freud denied the crowd any unconscious of its own (p. 378). Instead, he maintained that the unconscious itself operated in two ways: structurally, as a repressed historical force, and metaphorically, where the crowd merely amplifies what already exists in the individual unconscious (p. 378). It manifests frustrated desires (p. 381), with its violence paralleling teenage hysteria around musicians (p. 383). Ultimately, Freud displaced the crowd's power onto the leader and its creativity onto the poet, enclosing collective intensity within bourgeois institutions (p. 383).

Here, one should not throw the baby out with the bathwater, and if Freud, despite himself, is not the discoverer of a universal truth of the human species nor did he 'invent' the unconscious (as would go the vulgar narrative), he has the precise value of describing the ideal logic (ideology) of bourgeois society: his ideas reveal the limits and constraints of conceiving society, agency, substance and political action in a concept of an essential and finite individual. One further points is worth noting from Jodi Dean's account of the enclosure of the subject. She inverts Althusser's famous formula by saying that it is not the case that ideology interpellates the individual as subject, rather, capitalism interpellates subjects as individuals.

For our critique, it is not enough to follow Marcuse's approach, which highlights Freud's prejudices; rather, we must break from Freud's conception of subjectivity as an 'energy' centred on the individual. Freud's framework mirrors a geocentric model, where all social forces orbit around the 'genius' figure. This self-centred perspective underpins Freud's theory of narcissism, where the unconscious drives the ego through primitive self-love, projecting itself onto external objects only when serving self-interest (Freud 1957, pp. 76–77). Lacan (1949) critiques Freud's mirror-stage theory, introducing the 'doubled mirror'; later, he shifts towards topological models, recognizing a dialectical tension between drive and signifier (Vanheule 2011, p. 8), yet still centres subjectivity within the individual. In contrast, we need a Copernican revolution (Fig. 7.2), that pictures the whole system from above. Althusser (2000) sees the doubled mirror not as self-reflection but as a mechanism projecting subjectivity as an ideological construct rather than an inherent reality. Subjectivity is the Eter. Arguably, Simondon's scheme provides a theoretical framework with which to overcome the duality between individual and collective, thus supporting a 'Copernican' revolution in the conception of individual and collective. In this sense, being is not seen as substance—a centre—but as a system in system.

While analysing the multiple layers of the 'production of subjectivity', Guattari (1995, p. 22) proposed 'to decentre the question of the subject onto the question of subjectivity'. Subjectivity would be an ensemble of delimiting and self-referencing conditions, formed by an ecology of apparatuses and incorporeal universes of references, institutions, spatialities and ethic-aesthetic paradigms (Guattari 1995, p. 9). As such, Guattari argues that an approach to 'subjectivity' should be less of a scientific paradigm and more of a multi-faceted cartography of cognitive references, myths, rituals, symptoms, anguish, affect, inhibitions, drives, family relations and neighbourhood identities (Guattari 1995, pp. 11–12).

Thus, he argues for a map with a series of musical metaphors of existential orchestrations, which are precarious partial objects, such as harmonies, polyphonies, counterpoints, rhythms, and aesthetic subjectivities (Guattari 1995, p. 18). For him, subjectivity is pre-personal, polyphonic, collective and machinic; thus, it is not centred on the individual (Guattari 1995, p. 21). Nevertheless, Guattari is struggling against structuralism, using its theoretical apparatus to build a complex theory rather than a theory of that complexity. The result is the creation of complex structures, with multiple elements layered in partial levels and made of complex refrains, with temporary (yet precise) consistencies, interfaced by non-discursive virtualities, and so on (see especially Guattari 1995, pp. 23 and 27). To use his musical metaphor against his argument, such a theory could explain the complex functioning of an orchestra but not the dynamics of jazz improvisation.

One Last Map to Overcoming the Subject-Territory Dichotomy

Understanding how subjects engage with and transform reality requires examining different conceptions of subjectivity and its relationship with society. This section explores four key metaphors—the ideological individual, the interindividual, the prince, and the transindividual—each illustrating different ways in which subjectivity is structured and understood. These metaphors offer insight into how agency is framed within power structures, social formations, and collective dynamics, ultimately leading to a more nuanced perspective on the processes of individuation and transformation.

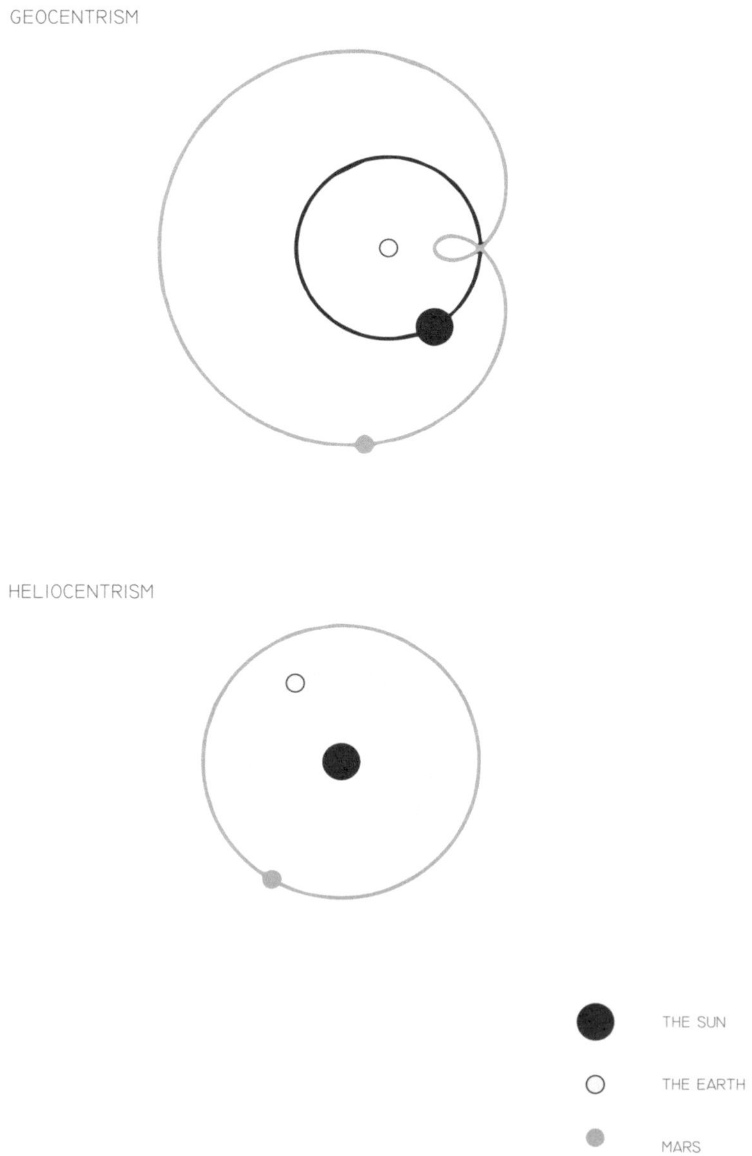

Fig. 7.2 Orbit of planets before and after the Copernican revolution. (Source: the author)

The ideological individual is the simplest and most widely accepted metaphor. It assumes a neutral field where individuals compete based on personal merit. This perspective legitimizes hierarchies by attributing success to talent and failure to personal flaws, reinforcing the notion of a leader at the top and others below due to their deficiencies (Fig. 7.3). However, this metaphor fails to acknowledge that social conditions and the urban field are shaped by power struggles, institutional structures, and historical contexts. It ignores the ways in which dominant institutions establish the rules of the game, ensuring their own reproduction and hegemony. The ideological framework thus obscures the actual mechanisms through which power operates, creating the illusion of a fair and neutral system.

Georg Simmel's concept of interindividuality challenges the isolated individual by emphasizing how subjects are formed at the intersection of different social circles. These circles—such as class, gender, profession, and culture—overlap to produce complex individualities, where knowledge and identities are formed between individuals in interaction (Fig. 7.4). This metaphor shifts the focus from an autonomous individual to a relational and networked conception of subjectivity, recognizing that ideas and values emerge through collective exchange rather than in isolated geniuses. However, this model still assumes an inner-centred individual, although shaped by external forces. It does not fully explain how individuals activate or embody different aspects of their identity at different moments, nor does it address the formation of collective subjects that transcend individual consciousness.

Fig. 7.3 The ideological individual: the myth of meritocracy. (Source: the author)

The prince, as theorized by Machiavelli and later critically examined by Althusser, represents a public image that condenses political power and collective desires. This metaphor understands leadership not as natural but as an articulated social construct, shaped by strategic manoeuvres and representation. The prince functions as a social mechanism, absorbing collective power while simultaneously enforcing control over it (Fig. 7.5). While this model reveals how ideological power operates, it also limits agency. Individuals subjected to this framework often feel trapped within a system they cannot change, as seen in The Wire (2003), where characters remain entangled in the structures of capitalism despite their efforts to resist. The prince metaphor, therefore, exposes the fetish of leadership but does not offer an alternative vision for collective empowerment beyond hierarchical power structures.

Alternativelly, the concept of transindividuality in Simondon moves beyond fixed notions of the individual. Instead of treating subjectivity as a stable entity, it focuses on the process of individuation, where individuals emerge through their relationships with others and their environment. Rather than being pre-defined, subjects are in constant becoming, shaped through interactions, social transformations, and collective practices. This metaphor offers the most open-ended and dynamic perspective, emphasizing that subjectivity is neither fully determined by external

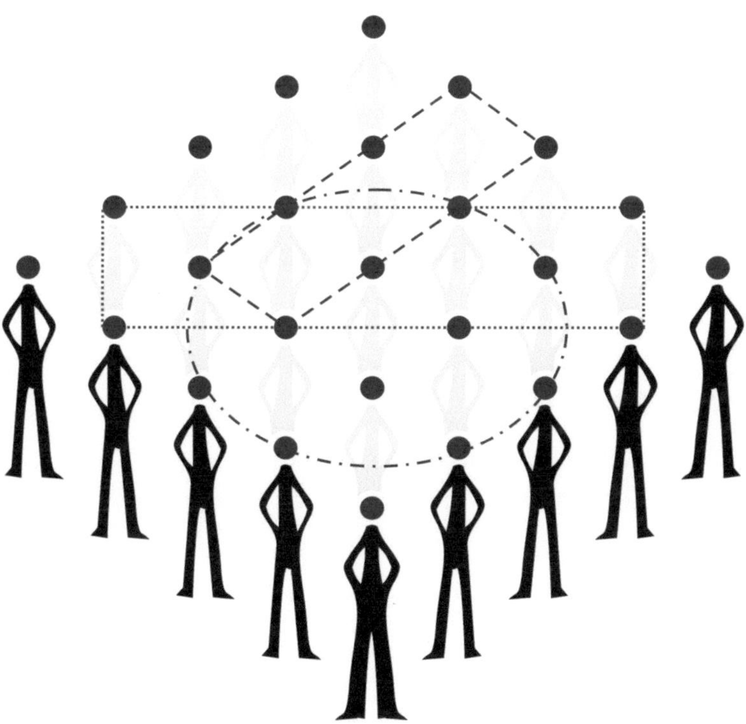

Fig. 7.4 Intersubjectivity: identity as an intersection of social circles. (Source: the author (based on—but not identical to—Simmel, *in* Stoetzler 2016))

Fig. 7.5 The Prince: power, representation, and social control (based on—but not identical to—Machiavelli 2008; Althusser 2000)

structures nor purely self-contained. Instead, it is a dialectical process that continuously unfolds, creating new possibilities for collective action and spatial transformation.

Each of these metaphors offers a different perspective on subject-society relationships, exposing the ideological underpinnings of agency and social positioning. While the ideological individual and interindividual models remain constrained by notions of self-contained identity, the prince metaphor unveils the mechanisms of power while still reinforcing centralized structures. The transindividual, however, presents a radical shift, breaking from static representations and embracing becoming, transformation, and collective agency. By rethinking subjectivity through this lens, new pathways emerge for understanding how subjects and collectives can actively shape and transform their social and spatial conditions.

7.2 Another Narrative of the Chain of Events: Transindividual Becomings

There is a gap between how production is actually done and the narratives that explains creation and mobilizes people to work—a gap that conceals the fundamental contradiction between the collective character of cultural production and the rhetoric of the 'genius'. In an effort to deconstruct this contradiction, we aim, in the discussion below, to critically address the hegemonic forms of narratives so we can transform the way we conceive the subjects of and their relation to the reproduction (or transformation) of social relations.

How does the reification of the architecture of the city result in the reification of subjects themselves? A similar difficulty has been depicted in the controversial movie Exit Through the Gift Shop (2010), by Banksy. On the one hand, the movie depicts an aspiring artist—a friend of Banksy—who uses a series of artifices to become famous, even without any apparent talent or deep understanding of the philosophical implications of that practice. On the other hand, what lies below the narrative is the fact that the atelier of Banksy and the field structure that makes him a star artist and his atelier operating with a team of hired artists are the same strategy used by the protagonist. The paradox rests, therefore, not so much in the fake artist pretending to be radical (while using the 'system of art' in a fetishist way) but in the artist who is willing to be radical but finds himself trapped in the reproduction of the same reified practice.

Our argument is not simply that the field encloses the artists and renders them subject to its heritage. Rather, we must deconstruct the very notion of an agent positioned against a field: a subject confronting an object, a radical soul trapped in a body, a consciousness separated from the unconscious. More broadly, we seek to uncover the theatre in which collective subjectivities are always in a process of becoming.

Hegemonic Narratives

Emerging at the end of the nineteenth century, the 'biography' genre became fundamental for the individualistic subject of capitalist society. Kracauer (1995 [a collection of texts ranging from 1920 to 1933], pp. 101–106) argues biography to be the 'Art Form of the New Bourgeoisie'.[1] For him, the novel of the previous century, in which individual unity was immersed in an overall context, was replaced by (and condensed in) the history of highly visible heroes. As the 'actual life' of individuals gives a sense of 'certainty' to historical veracity, facts become crystallised, making

[1] In a digital media era, perhaps this art form is not decaying; rather, it is imploding into the form of 'profiles': abbreviated and simulated representations in the shapes of avatars.

history the ultimate result of individual actions. For Kracauer, this is the ultimate form of 'evasion' from the masses and the collective character of history.

Beneath that, the imagination of agency became trapped in a deeper conception of history based in the chronicles of 'princes' (as we saw in Machiavelli by Althusser 2000). The intellectual operation where the 'prince' becomes 'the subject' of history, many scholars have ultimately reinforced the abstract device of the 'prince' because the critique was centred on these representations (be they architects or iconic buildings) conducting the evolution of the history of the city.[2]

In critical theory, attempts to re-imagine the subject of history have revisited texts such as Marx's *The Eighteenth Brumaire of Louis Napoleon* and *The Civil War in France*. In those texts, Marx uses his historical method: in the former, he demonstrates the gradual formation of mechanisms of control in the formation of modes of production, and in the later, he aims to demonstrate the Paris Commune as a critical moment of increasing contradictions in society. In the *Communist Manifesto*, Marx and Engels n.d. [1848]) set out his famous concept of history as the eternal struggle between classes. Thus, class becomes the central subject of history. This implies that 'class consciousness' is a key element for historical change, as Lukács (1971) observed.

Thus, a theoretical problem would soon emerge: that of organisation (Lukacs n.d. [1924]). This problem was first developed by Lenin (1961 [originally written in 1902]) in the pamphlet *What is to be done?* His solution was to conceive the Communist Party as the 'vanguard' of the proletariat, thus entitled to act on the behalf of that class. Nevertheless, Freire (1996) would famously assert that those who are oppressed by society could not enter the struggle for freedom in the condition of objects and only later become fully entitled as human beings, otherwise the struggle would be lost from the start. It is in this sense that Freire proposes a pedagogical strategy centred in the development of subjects who are critically conscious of their place in society.

In this discussion, the important point to retain is what Amir Djalali (2014) has identified as the 'common' of architecture, wherein architectural knowledge is produced by the entire field, whose origin is collective in nature. Although this production is collective, its products are captured by social apparatuses. For Jason Read (2010, p. 123) the political problem arrives because this subjectivity is 'reified' (framed/moulded/distorted) by abstract social machines and structured by a 'fetishisation in which the qualities and attributes of social existence are attributed to a thing'. Thus, 'Society itself exists as a fetish' (Read 2010, p. 125) and therefore Read asserts that, as Marx had put it before, 'production not only creates an object for the subject, but a subject for the object' (p. 120). For Read (2010, p. 114), this idea is

[2]For instance, Tafuri's (2006) last work was 'Interpreting the Renaissance: Princes, Cities, Architect,' which deployed many interesting interpretative tools to consider the interactions among those actors.

> often seen as tantamount to a denial of political agency altogether, to the assertion that everything is an effect of power, that agency and action cannot exist. What I would like to propose is that far from being a theoretical dead end for politics the production of subjectivity is the condition for its renewal (…) It is a matter of articulating this common, the unrepresentable transindividual collectivity, against the conditions and practices that conceal it (…) not a matter of looking for 'the subject' (Read 2010, p. 130).

The historical imagination of agency has long been confined within the biographical and princely narratives that reduce collective social forces to the actions of exceptional individuals, obscuring the collective forces that shape reality. This perspective reinforces hierarchy, portraying success as personal merit while ignoring structural conditions. In contrast, history is a dynamic process of struggle, shaped by shared consciousness and collective agency. However, dominant systems reify subjectivity, turning social relations into fixed structures that appear inevitable. Breaking from this illusion requires recognizing subjectivity as an evolving, transindividual process—one that emerges through relations, enabling new possibilities for collective action and transformative spatial practices. However, how can a collective subjectivity possibly be mobilised?

Another Chain of Events

For Alain Badiou (2012), an 'event' is the moment when a new political truth is collectively accepted. Thus, an event is the emergence of a truth into being. The event is a process of collective subjectivation, in which a group of people accept a new appearance of a given social condition and mobilise their will to act. In this sense, an event is similar to a work of art because it brings about a new aesthetic of things.

In this process of event as an art production, Zizek (2014, pp. 136–148) argues there is a reversal of contingency into necessity: a master-signifier of the past is changed in order to make the present its inevitable result. In this sense, the unity imposed by a work of art re-appropriates the past. This is not simply illusory, nor simply imaginary, as it establishes a new social reality by symbolically restructuring the entire field (even though there is no new content, everything might seem different) and retroactively changes the past chain of causalities that brought us to the present. Thus, the creativity of the work of art as event is to make visible a chain of causality in a process of transformation.

Badiou calls this an 'idea' that assembles collective participation in a singular political process in the form of a historical decision. It is retroactively projected onto the symbolic movement of history, giving a name to those who had no say (Badiou 2012, pp. 8–9). An event is the process by which a social necessity appears, rupturing 'the normal order of bodies and languages as it exists for any particular situation' (Badiou 2012, p. 6), becoming the affirmation of something that was, up until then, 'impossible'.

However, this conception of an event is not one of a single point in time (an instant), nor is it a variation of something like a function or an activity (such as in

Tschumi 1994). Rather, as Zizek (2014) draws upon Badiou, an event emerges when reality is reframed by a sudden moment of truth (as in an insight), or by a subject who recognises how his cogito builds his own existence, or when a mistake's incongruence with reality reveals the abyss of everyday madness, or when the imaginary realm of social representation is unmasked and recreated. Thus, events transform both subject and reality. For Feltham (2006), commenting on Badiou, 'The "and" of "being and event" is thus up to the subject: it's open' (Feltham 2006, p. xxxi).

Beyond Agency

It is important to recognise that the research project entitled *Spatial Agency: Other ways of Doing Architecture* (Till et al. 2011) has made important contributions to new understandings of the field of architecture and of how architects contribute dialectically to the production of space. Most remarkable is how this project presents the different practices, discarding any linear or compartmentalised narrative and rupturing with the notion of linear evolution in the field. Rather than presenting a series of affiliations and schools, the report presents practices in alphabetical order, simultaneously eliminating both hierarchy and precedence. Practices are addressed as a network of interconnected yet individual contributions.

This approach is developed, fundamentally, by merging Lefebvre's (1991b) notion that space should be understood as a social product, Latour's (2005) theory that actors should be understood within a network, and Giddens' (1984) idea that 'agency and structure should be understood as a duality, two linked but separately identifiable conditions'. The 'Spatial Agency' book (Till et al. 2011) allegorically starts with a 'half-joke' made by Latour: 'There are four things that do not work with the Actor-Network-Theory... the word actor, the word network, the word theory and the hyphen'.

Latour's (1999b) intention was to stress his proposal as a method of reconstructing how actors build their activities rather than to formulate a new social theory (which he sees as 'grand-narratives'). Nevertheless, he ultimately concedes that his theory fails to take into account the 'pressure' of epistemology (what is 'out there') and of psychology (what is 'in there') (Latour 1999b, p. 23). Arguably, to advance beyond the ideas of an 'actor', of a 'network' and of a 'hyphen', it would be necessary to overcome the long-lasting structuralist model of 'langue-parole' (the logical origin of 'network-actor', in this case).[3] Furthermore, it could be argued that the proto-post-structuralist agency theory of Giddens (1984) is, in his words, an attempt to overcome this 'dualism' by means of a 'duality' (sic), thus—more or less—delivering the same partial self-criticism as Latour.

---

[3]Arguably, what Latour is recognising (without saying so) is the socio-political instance of the production of subjectivity.

Pierre Bourdieu (2009, p. 39) tries to overcome this paradox by considering the exchange of 'the model of reality for the reality of the model'. Thus, he proposes that although the structure is a social representation, it has actual impact. This is a relevant point, and therefore, the analysis of the structure of the field matters. Nevertheless, this account does not illuminate the way in which this structure is actually internalised in individual subjectivities, nor does it allow us to build a form of action that moves beyond the necessity of a 'personal' actor who provides the creative shift in the structure. The solution of this paradox will demand the reframing of the problem altogether.

This step could not be a caricature of post-structuralism, which bases itself in a complexification of the structure, critiquing and deconstructing it endlessly, aiming to provide the utmost model of the continuum of reality. Once again, we should remember that the aim of this book is much more modest. Rather than providing a new 1:1 cartography of the absolute complexity of reality and its cognition, we aim to produce a map that allows us to navigate it, to build shortcuts, to create deviations and, hopefully, to allow readers to conceive possible detours around these paradoxes.

The fundamental problem of the idea of 'agency' is the assumption of an 'autonomous' subject, acting with free will in a structure (that can be neutral or not). Nonetheless, to move beyond the concept of agency does not aim to deny the possibility of action; rather, it aims to engage in how social structures frame possible choices, condition alternatives, foment drives, and induce behaviours by expectation; furthermore, to engage in how subjects are (from the start) subjected to a past that goes beyond individuals and how that awareness is the first step to deconstruct and reconstruct non-actualized potentialities in transubjective—collective—events.

7.3 Towards a Political Ontogenesis of Space

For Rancière, there is a poetical struggle in the production of history, where those in positions of power aim to make history the result of their own actions, thus silencing the concrete history experienced by others. Thus, aesthetics and visibility are key issues in critical analysis. For Rancière, aesthetics is a political act, which has nothing to do 'with aesthetic utopia, with a certain idea of artistic radicality and its capacity to perform an absolute transformation of the conditions of collective existence' (Rancière 2009, p. 19). Rather, 'what links the practice of art to the question of the common is the constitution, at once material and symbolic, of a specific space-time, of a suspension with respect to the ordinary forms of sensory experience' (Rancière 2009, p. 23). Thus, art is not political because of the way it represents society but rather because of the distance it creates from society and the type of space it has enabled as a result. A political aesthetic elicits awareness of the mechanisms of domination, thus turning a passive 'spectator into a conscious agent' (Rancière 2009, p. 45).

Kracauer (1995), in his analysis of the 'Hotel Lobby,' adopts a different strategy. Instead of developing a rigorous analysis behind the 'discourse' of famous architects' buildings—the magician's trick seen from behind the scenes—he aims to investigate how, in the process of aesthetic production, by aiming to produce a totality and ultimately failing to do so, the aesthetical act—due to its contextual condition—ends up maintaining this totality as the veiled frame of its appearances. In this sense, what the object obscures is more important than what it reveals (just as Tafuri's blind spots reveal more about the discipline than do his own negative critiques). In this sense, his analysis, for instance, uses the detective novel as a means to show society its 'own face in a purer way,' making it possible to reveal 'a whole out of the blindly scattered' (Kracauer 1995, p. 175). In this sense, there is no logical justification for focusing on examples of 'star' architects to reveal the contradictions of the field. In Kracauer's words, 'For no matter how insignificant the existential power that gives rise to the artistic formation may be, it always infuses the muddled material with intentions that help it become transparent' (Kracauer 1995, p. 174). This reveals a double dialectic. On the one hand, both 'high art' and 'mass ornament' deliver, as by-products of their aesthetics, precisely what the artist and the culture cannot entirely grasp or see.

It is in this sense that Walter Benjamin (n.d. [1940]) famously proposed the task of brushing 'history against the grain'. For him, our awareness of the present is moulded by those who have been victorious; thus, we should make an effort to revive the views of those who have been defeated. This is an approach to history as a situated assessment of contradictions empirically observed, followed by a contextualised excavation of its origins, followed by a reframing of the problem. In our case, this approach shows how, despite all its rhetoric of innovation, creativity and revolutionary conception, the urban spaces in London concretely operate as an instrument of reproducing social relations.

Nonetheless, the possibility of collective individuations is possible because the individual is just a provisional actualisation of a shared field of preindividualities, thus formed of internal 'disparations'. An event is necessary to produce a mediation, or modulation, able to interconnect un-actualised fields of possibilities. This collective dimension of individuality is what Simondon calls the transindividual, expressed in the diagram above by the letter (T). In this sense, Toscano (2006, p. 153) explains that the preindividual is not an a priori, just as transubjectivity is not an a posteriori. Rather, this allagmatic expresses a systemic *a praesenti*: an informative and interactive process, at the same time bigger and smaller than the individual.

For Simondon (2013, pp. 298–303), the subject is the articulation of the three dimensions of individuality: preindividual, individuated and transindividual. His central point on the subject is that it emerges by means of a 'choice'. This choice creates an actualisation of a part of the internal preindividual that had not been individuated previously. However, it can only emerge by means of a collective. This choice is thus a process of structuration inside the subject but directed towards other subjects. For Simondon, this implies that the subject is more than the individual. Nonetheless, Simondon does not include in his reasoning how ideology, society, hierarchy, social struggles and conflicts (i.e., politics) interfere in the production of preindividualities and transindividualities (be they disciplined or radical).

Toscano (2007) even suggests that one could see an 'anti-political tendency' in Simondon's writings, if one takes, for politics, terms such as 'sovereign' and 'democratic representation'. Nevertheless, Toscano asserts that it provides fertile ground to think about 'pre-revolutionary' conditions and the emergence of resolutions of internal conflicts. In this sense, one would need to think of politics as a process of philosophical 'torsions' of metastable individualisations that occur through the production of events. As in the crystallisation of water in metastable conditions, an event is ignited by means of the insertion (invention) of a 'structuring germ', able to mediate and create new transubjective individuations.

Although the concept of transubjectivity enables the understanding of the emergence of transformation, the theory of Simondon does not advance in the political behind these transformations: the struggle for the emergence of new subjectivities; and the conflict between different 'pre' and 'trans' individualities. Therefore, beyond acknowledging the production of individualisation, we will need to discuss the (re)production of presubjectivities and transubjectivities themselves. For that, we will return to the dialectical psychology of Vygotsky (n.d.-b [1930]) so as to inform this discussion politically. If the production of transubjectivities is a process/event, to advance further, we will have to approach how the 'subjective senses' are produced. In this way, we can politicise the debate over transubjectivity.

González Rey (2007) rescues the concept of 'subjective sense' and that of 'units of subjectivities' as key concepts in the late work of Lev Vygotsky. These concepts are concerned with the production of the social dynamic configurations of the collective psyche. Against the understanding of the human mind as a blend of standardised traits, Vygotsky was concerned with the dialectical production of consciousness. For that, he expands the Freudian concept of transference to better comprehend external influences on personal behaviour.

Nevertheless, González Rey (2007) asserts that Freudian psychoanalysis took the direction of proposing a universal subject, formed of essential libido energy. Thus, by proposing an abyssal division between psyche and culture, modern psychology ended up reifying a specific cultural condition into its theory. The use of the term subjectivity by Soviet Psychology aimed to break down this abysm. Subjectivity does not aim to encompass all the phenomena of causalities or a description of substances; rather, it aims to create an 'intelligibility of this reality'. For this reason, Vygotsky does not eliminate the use of the term 'unconsciousness' but cleanses it of the 'internal universal drive'.

In this sense, González Rey (2007) asserts that Vygotsky's concept of 'subjective sense' aims to address the configuration of subjectivity resulting from lived experiences, historical background and contexts of life, including the relationship among activities, memories and fantasies in a network of what is symbolic and emotional. In this sense, 'The psyche becomes a symbolic production within cultural scenarios, yet it is not reduced to a symbolic phenomenon' (González Rey 2007, p. 6). In this way, the sense of a word can only be understood in the context of a phrase, a phrase in a book, a book in the work of an author, an author in the context of a nation, a nation in the context of the world. Thus, subjective senses are integrations between cognitive and affective processes, forming a 'dynamic system of senses'.

> The social becomes subjective not because of internalization, but by sense production related to living experience... The subjective sense is not confined to being intrapsychical. It always exists in the continuous tension between already organized subjective configurations, which are always under pressure from social subjectivity, and the subject's action (González Rey 2007, p. 9).

Subjective senses are the units of a dynamic system of subjectivities, and they are defined by how different social spaces are interrelated. These senses create internal contradictions, and an important source is the 'Social representations and discourses from which the social evaluation of people is produced' (González Rey 2007, p. 11).

Finally, we arrive at what Toscano (2006, 153) considers the fundamental concept of Simondon's method: 'transduction'. This concept opposes traditional reduction, deduction and induction (see Fig. 7.6 below). Transduction is conceived as the process of resolving internal contradictions and energy conflicts inside an

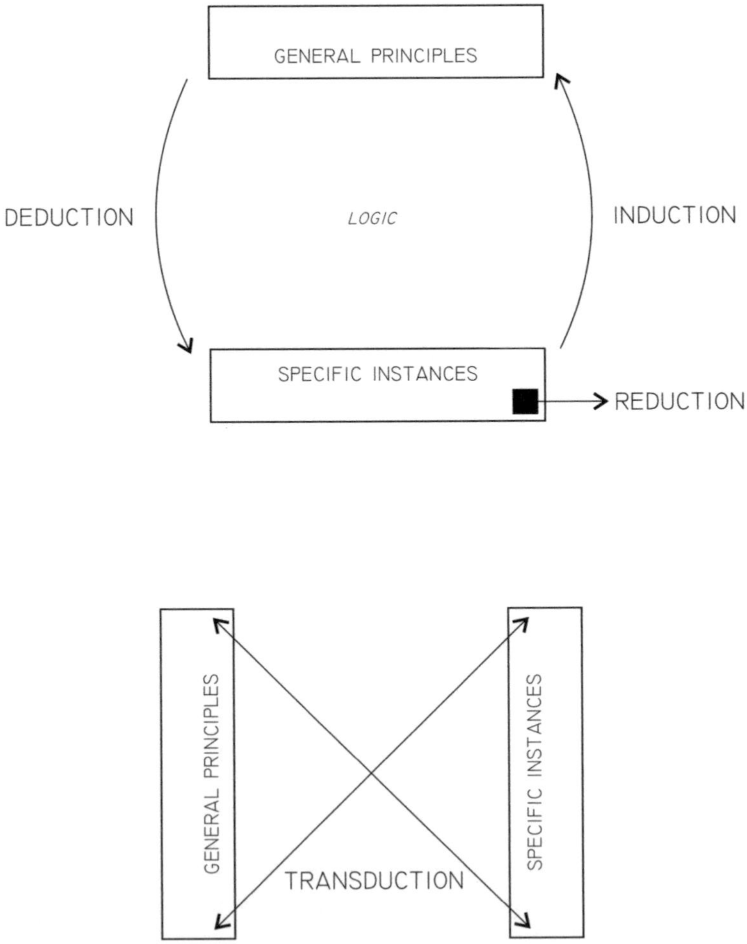

Fig. 7.6 Deduction, induction, reduction, transduction. (Source: the author (based on—but not equal to—Simondon 2013, and Toscano 2006))

individual. In this sense, information is not a form of communication between two separate beings; rather, it is the production of a 'mediation', of an element that potentially connects two internal dimensions, thus bringing into being what was not realised in the individual (previously existing only as the preindividual). It is in this sense that Simondon (2013, p. 170) asserts that an individual is not a virtual (in the Aristotelian sense) that actualises itself (always in the same way); rather, it is a transduction that performs a real transformation of the possibilities of the system.

To highlight this distinction, Combes (2013) differentiates virtuality from potentiality. For Simondon (2013, pp. 68–69), individuation must be thought relationally, introducing the concept of a system with potential energy—the capacity to transform energy into form, or matter, i.e. work. As Combes (2013, p. 50) asserts, Simondon shifts the Kantian question from '*What is man?*' to '*How much potential does a human have to go beyond itself?*' Extending this logic, the question is not *What is the city?* but rather *How much potential does the city have to go beyond itself?*

This leads us to a crucial challenge: how to conceive the architecture of the city not as a static order but as a transformative process? Urban theory and epistemology often depart from a reified and disciplinary stance, reinforcing power structures that reproduce subjectivities rather than enabling agency. As this chapter has mapped, disciplinary mechanisms function as structured mediations, ensuring that spatial practices reflect and reproduce specific subject positions, enclosing stakeholders in phantasmagorical narratives. Urban design and architecture, rather than merely designing the built environment, actively conditions modes of perception, habits, and subjectivity—a process often concealed by the field's own rhetoric of innovation and authorship.

This recognition underscores the core paradox of architecture, having a role in shaping society while being unable to transform it at a structural level. The problem is not merely that radical design practices fail but that the very ideological structure of spatial production reproduces itself. The challenge, then, is not to create new stylistic solutions or utopian models but to break from the cycle of disciplinary enclosure—not through naïve gestures of resistance but through critical interventions that expose and reframe the ideological structures embedded in spatial production.

Thus, the aim of this book was never to propose a blueprint for 'new practices', but to offer reflexive tools for recognising the role of urban interventions in the reproduction of social relations. If Marx (n.d. [1867]) described *Capital* as "without question the most terrible missile that has yet been hurled at the heads of the bourgeoisie," this book instead seeks to ignite transformation from within the consciousness of its readers. It operates as a cognitive map, exposing how urban structures sustain neoliberal subjectivities, and how design practices, even when radical in intention, often fall back into ideological complicity.

So, why do radical design practices so often fail? Because transformation demands more than technical solutions, stylistic alternatives, or new methodologies. It requires a rupture in the epistemic field: a reconfiguration of how we think, teach, and practice urban space. This means moving beyond fetishized deadlocks towards a cartography of power relations—one that not only denounces the status quo but interrogates how subjectivity itself is reproduced within spatial regimes. Only then can we begin to articulate new transubjective and collective agencies for a political ontogenesis of space. In this light, the city's potential does not reside in prescriptive models, nor in the urban form itself, but in strategic acts of de-reifying and reframing both material and immaterial mediations, opening pathways for a constant process of collective liberation.

References

Adams RE (2014) Notes from the resilient city. Log 32, pp 126–139
Adorno TW (1951) Minima Moralia. Suhrkamp, Frankfurt am Main
Adorno T (2004) Negative dialectics. Taylor & Francis e-Library, London
Adorno T, Horkheimer M (1996) Conceito de Iluminismo. Nova Cultural, São Paulo
Adu A, Horton H (2025) Sadiq Khan to announce plans to build houses on London green belt. The Guardian, 9 May. https://www.theguardian.com/uk-news/2025/may/09/sadiq-khan-to-announce-plans-to-build-houses-on-london-green-belt. Accessed 25 May 2025
Alberti LB (1988) De Re Aedificatoria. MIT Press, Cambridge, English translation
Alexander C (1977) A pattern language. Oxford University Press, New York
Althusser L (1971) Lenin and philosophy and other essays. Monthly Review Press, New York
Althusser L (2000) Machiavelli and us. Verso, London
Anonymous (2014) Interview to the author
Appleton J (2014) The end of public space. The Architectural Review, 15 January. Accessed 27 Jan 2015
Arbuckle Industries (2015) Mary Woods architecture extras interview. https://www.youtube.com/watch?v=lni9hFemEZQ. Accessed 8 Apr 2016
Arendt H (2006) On revolution. Penguin Books, New York
Aureli PV (2008) The project of autonomy. Princeton Architectural Press, New York
Aureli PV (2013) Less is Enough – On Architecture and Asceticism. Strelka Press, Moscow
Badiou A (2012) Being and event. Continuum, London
Bartocci U (1999) Albert Einstein e Olinto De Pretto. Editore Andromeda, Bologna
Batty D (2016) How London's booming "butler class" takes care of the wealthy elite. The Guardian, 12 March. https://www.theguardian.com/business/2016/mar/12/family-office-private-wealth-funds. Accessed 16 Mar 2016
Baudelaire C (1995) The painter of modern life and other essays. Phaidon Press, London
Baudelaire C (1996) Pequenos poemas em prosa. Editora da UFSC, Florianópolis
Baudrillard J (1988) America. Verso, London
Baudrillard J (1991) Simulacros e Simulação. Relógio D'água Editores, Lisboa
Bauman Z (2010) Socialism: the active utopia. Routledge, London
Bellis M (2016) Thomas Edison's Muckers. http://inventors.about.com/od/famousinventions/fl/Thomas-Edisons-Muckers.htm. Accessed 30 Oct 2016
Benjamin W (2008) The work of art in the age of its technological reproducibility, and other writings on media. Harvard University Press, Cambridge
Benjamin W (n.d. [1940]) On the concept of history. https://www.marxists.org/reference/archive/benjamin/1940/history.htm. Accessed 12 Jan 2015

C. V. De Lima Amaral, *The Urban Reproduction of Subjectivities*,
SpringerBriefs in Geography, https://doi.org/10.1007/978-3-032-01465-8

Benn S, Gauss G (1983) The public and the private. In: Benn S, Gauss G (eds) Public and private in social life. S. St. Martin's Press, New York, pp 3–30
Berardi F (2011) In: Genosko G, Thoburn N (eds) After the future. AK Press, Oakland
Berman M (1990) All that is solid melts into air, 5th edn. Verso, London
Better Elephant (2013) Heygate state dislocation. Community Panflet
Bishop C (2006) The social turn. Artforum 44(6):178–183
Bishop C (2011) Participation and spectacle [Lecture for creative time's living as form]. Cooper Union, New York. http://dieklaumichshow.org/pdfs/Bishop.pdf. Accessed 14 Jun 2015
Boffey D (2023) How will recent and future legislation affect the right to protest in the UK? The Guardian, 13 November. https://www.theguardian.com/world/2023/nov/13/how-will-recent-and-future-legislation-affect-the-right-to-protest-in-the-uk. Accessed 20 Jan 2025
Booth R (2015) Barnet council 'engaged in social cleansing' at West Hendon estate. The Guardian, 20 January. https://www.theguardian.com/society/2015/jan/20/barnet-council-social-cleansing-barratt-homes-west-hendon. Accessed 13 Mar 2015
Borough Council of Waltham Forest (2013) Wood street area action plan, planning policy document. https://www.walthamforest.gov.uk/sites/default/files/2021-11/Wood%20Street%20AAP%20-%20PO%20Stage%20Web%20version-LR.pdf. Accessed 30 Sep 2014
Bourdieu P (1996a) The rules of art. Polity Press, Cambridge
Bourdieu P (1996b) Distinction. Harvard University Press, Cambridge
Bourdieu P (1996c) The state nobility. Stanford University Press, Stanford
Bourdieu P (2009) The logic of practice. Polity Press, Cambridge
Bourriaud N (2002a) Relational aesthetics. Les Presse Du Reel, Paris
Bourriaud N (2002b) Conversations. Interview with Nicolas Bourriaud and Karen Moss. Interviewed by Ella Delaney for Stretcher, 18 October. http://www.stretcher.org/features/nicolas_bourriaud_and_karen_moss/. Accessed 20 Feb 2015
Brouillette S (2009) Creative labor. Mediat J Marxist Literary Group 24(2):140–149
Brown JB (2014) Learning theories for live projects. In: Harriss H, Widder L (eds) Architecture live projects—pedagogy into practice. Routledge, London, pp 18–23
Butler J (1988) Performative acts and gender constitution. Theatr J 40(4):519–531
Cabral Filho J, Baltazar A (2010) Magic beyond ignorance: virtualizing the black box. In: de Arte Digital F (ed) Essays. Instituto Cidades Criativas, Rio de Janeiro, pp 12–23
Cameron D (2016) I've put the bulldozing of sink estates at the heart of turnaround Britain. The Sunday Times Columns, 10 January. http://www.thesundaytimes.co.uk/sto/comment/columns/article1654318.ece. Accessed 17 Feb 2016
Carlson WB (1988) Thomas Edison as a manager of R&D. IEEE Technol Soc Mag 7(4):4–12
Castells M (2002) A sociedade em rede, vol 1, 6th edn. Paz e Terra, Sao Paulo
Catling CS (2014) Damned if you do, damned if you don't. Architecture Review, 22 September. https://www.architectural-review.com/rethink/viewpoints/damned-if-you-do-damned-if-you-dont-what-is-the-moral-duty-of-the-architect/8669956.article. Accessed 15 Dec 2014
Chandler A (2014) Building is also a verb. In: Harriss H, Widder L (eds) Architecture live projects—pedagogy into practice. Routledge, London, pp 65–71
Chomsky N (2003) Pirates and emperors, old and new. South End Press, Cambridge
Combes M (2013) Gilbert Simondon and the philosophy of the transindividual. MIT Press, Cambridge
Comte A (n.d. [1848]) Discurso Preliminar sobre o Espírito Positivo. Editorial Ridendo Castigat Mores. www.jahr.org (Downloaded: 15 January 2008)
Contieri RP (2005) Einstein: annus mirabilis. Exacta 3:29–33
Crary J (2013) 24/7—late capitalism and the ends of sleep. Verso, London
Curtis A (2002) The century of the self [Documentary]. BBC, London
Darrigol O (2004) The mystery of the Einstein–Poincaré connection. Isis 95(4):614–626
Davies B (2006) Subjectification. Br J Sociol Educ 27(4):425–438

De Angelis M (2000) Marx's theory of primitive accumulation: a suggested reinterpretation. Working paper no. 29. Department of Economics, University of East Anglia, London. http://homepages.uel.ac.uk/M.DeAngelis/PRIMMACCA.htm
De Magalhães C (2010) Public space and the contracting-out of publicness. J Urban Des 15:559–574
Dean J (2016) Enclosing the subject. Polit Theory 44(3):363–393
Dees D (2012) Report on planning and housing compulsory purchase orders. http://www.compulsorypurchaseassociation.org/files/CPO-report-July-2012.pdf
Deleuze G (1992) Postscript on control societies. October 59:3–7
Department for Communities and Local Government (2005) Compulsory purchase guidance. The National Archives, London
Djalali A (2014) Common space. Unpublished PhD thesis, Delft University of Technology
Drageset O (2015) Negative energy explained by an imaginary speed of light. Phys Essays 28(1):92–94
Eisenman P (2006) O fim do classico. In: Nesbitt K (ed) Uma nova agenda para a arquitetura. Antologia teórica 1965-1995. Cosac Naify, São Paulo, pp 232–251
Engels F (n.d. [1880]) Do socialismo Utópico ao Socialismo Científico. http://www.marxists.org/portugues/marx/1880/sociutopsocicien/index.htm. Accessed: em 05/04/07
Erlanger S (2016) Grieving and going forward. The New York Times, 20 April. https://www.nytimes.com/design. Accessed 22 Apr 2016
Escobar A (1996) Constructing nature. In: Peet R, Watts M (eds) Liberation ecologies. Routledge, London
Exit Through the Gift Shop (2010) Directed by Banksy [Film]. Revolver Entertainment, London
Fairs M (2014) Designs of the year prize "about architecture rather than politics" says Design Museum director, Dezeen, 2nd July. http://www.dezeen.com/2014/07/02/design-museum-designs-of-the-year-prize-zaha-hadid-heydar-aliyev-centre-azerbaijan/. Accessed 10 Jul 2015
Feltham O (2006) Translator's preface. In: Badiou A (ed) Being and event. Continuum, London, pp xvii–xxxiii
Feyerabend P (1989) Contra o Método. 3a edição. F. Alvez ed, Rio de Janeiro
Florida R (2002) The rise of the creative class. Basic Books, New York
Flusser V (1985) Filosofia da Caixa Petra. Hucitec, São Paulo
Foerster HV (1972) Perception of the future and the future of perception. Instr Sci 1(1):31–43
Forst M (2024) Statement by the UN Special Rapporteur on Environmental Defenders under the Aarhus Convention, following his visit to the United Kingdom on 10–12 January 2024, UNECE. https://unece.org/sites/default/files/2024-01/Aarhus_SR_Env_Defenders_statement_following_visit_to_UK_10-12_Jan_2024.pdf. Accessed 12 May 2025
Foucault M (1979) Discipline and punish. Vintage Books, New York
Foucault M (1980) In: Gordon C (ed) Power/knowledge: selected interviews and other writings 1972–1977. Pantheon Books, New York
Foucault M (1984) Des espaces autres - Conférence au Cercle d'études architecturales, 14 mars 1967. In: Architecture, Mouvement, Continuité, no 5 (1984):46–49, 14 mars 1967. http://foucault.info/documents/heteroTopia/foucault.heteroTopia.fr.html. Accessed 2 Apr 2009
Foucault M (2005) Arqueologia das Ciências e História dos Sistemas de Pensamento - Ditos & Escritos, 2nd edn. Forense Universitária, Rio Janeiro
Frampton K (2000) História Crítica da Arquitetura Moderna. Martins Fontes, São Paulo
Freire P (1996) Pedagogy of the oppressed. Penguin, London
Freud S (1955 [1920]) Beyond the pleasure principle. In: Freud S (ed) The standard edition of the complete psychological works of Sigmund Freud, vol 18. The Hogarth Press, London, pp 1–64
Freud S (1957 [1914]) On narcissism: an introduction. In: Freud S (ed) The standard edition of the complete psychological works of Sigmund Freud—volume XIV (1914–1916). The Hogarth Press, London, pp 67–102
Friedland R, Zellman H (2007) The fellowship: the untold story of Frank Lloyd Wright and the Taliesin fellowship. Harper Collins, New York

Galloway A (2004) Protocol: how control exists after decentralization. The MIT Press, Cambridge
Galloway A, Thacker E (2007) The exploit. University of Minnesota Press, Minneapolis
Gayle D (2025) UK accused of undermining democratic rights with climate protest crackdown. The Guardian, 16 January. https://www.theguardian.com/environment/2025/jan/16/uk-accused-of-undermining-democratic-rights-with-climate-protest-crackdown. Accessed 12 May 2025
Giddens A (1984) The constitution of society. University of California Press, Berkeley
Gilbert J (2016) Culture, power, politics. Lecture at open school east, 17 May. University of East London, London
Gilens M, Page BI (2014) Testing theories of American politics. Perspect Polit 12(3):564–581
Gine J (2010) Einstein versus Lorentz and Poincare: open questions of credit. Phys Essays 23(1):92–96
GLA (2014) Housing in London 2014. Greater London Authority, London. https://www.london.gov.uk/sites/default/files/housing_in_london_2014_-_final.pdf. Accessed 14 Feb 2025
Glaser B, Strauss A (1967) The discovery of grounded theory. Aldine, Chicago
Gloster D (2014) Working margins, drawing lines. In: Harriss H, Widder L (eds) Architecture live projects. Routledge, London, pp 51–57
González Rey FL (2007) Social and individual subjectivity from an historical cultural standpoint. Crit Soc Stud 2:3–14
Guattari F (1995) Chaosmosis. Indiana University Press, Indianapolis
Guest G, MacQueen K, Namey E (2011) Applied thematic analysis. SAGE Publications, Thousand Oaks
Harris J (2025) It's crunch time for Starmer and Reeves: either build social housing or lose the next election. The Guardian, 8 June. https://www.theguardian.com/commentisfree/2025/jun/08/labour-housebuilding-plan-england. Accessed 19 Jun 2025
Harriss H, Widder L (2014) Architecture live projects. Routledge, London
Hart A (2025) We fought and beat the government in the courts because every Briton has the right to protest. The Guardian, 2 May. Available at: https://www.theguardian.com/commentisfree/2025/may/02/government-britain-protest-laws-appeal-courtlabour-democracy. Accessed 15 June 2025
Harvey D (1990) The condition of postmodernity. Blackwell, Cambridge
Harvey D (2004) The "new" imperialism. Social Regist 40:63–87
Harvey D (2010) A companion to Marx's capital. Verso, London
Harvey D, Whachsmuth D (2011) What is to be done? And who the hell is going to do it? In: Brenner N, Marcuse P, Mayer M (eds) Cities for people, not for profit. Taylor and Francis, Hoboken, pp 264–274
Hillier B (2007) Space is the machine. Space Syntax/UCL, London
Iveson K (2007) Publics and the city. Blackwell, Oxford
Jameson F (1981) The political unconscious. Cornell University Press, Ithaca
Jameson F (1990) Cognitive mapping. In: Nelson C, Grossberg L (eds) Marxism and the interpretation of culture. University of Illinois Press, Chicago, pp 347–360
Jameson F (1991) Postmodernism, or, the cultural logic of late capitalism. Verso, London
Jameson F (2000) Architecture and the critique of ideology. In: Hays KM (ed) Architecture theory since 1968. MIT Press, Cambridge, pp 441–461
Jameson F (2004) The politics of utopia. New Left Rev 25:35–54
Jobs (2013) Directed by Joshua Michael Stern [Film]. Open Road Films, Los Angeles
Johnson B (2009) A manifesto for public space: London's Great Outdoors—Mayor of London. Greater London Authority, London. www.london.gov.UK. Accessed 28 Nov 2013
Johnson B (2010) The London housing strategy. Greater London Authority, London. www.london.gov.UK. Accessed 7 Jun 2015
Johnson B (2013) Homes for London: the London Housing Strategy—draft for consultation. Greater London Authority, London. www.london.gov.UK. Accessed 12 Jul 2014
Johnson B (2014) Homes for London: the London Housing Strategy [approved document]. Greater London Authority, London. www.london.gov.uk. Accessed 8 Jan 2016

Jones L (2015) Interviewed by the author, 4th November, personal archive
Karthaus (2014) Construction week—wood street—sylalubus. In: ARM102: construction week—workshop 5. University of East London/ACE, unpublished
Karthaus R, Chandler A, Minton A (2013) 'Sylalubus', MAASD152.1: urban ecology. University of East London/ACE, unpublished
Kohn M (2004) Brave new neighborhoods: the privatization of public space. Routledge, New York
Koolhaas R (1994) Delirious New York. Monacelli Press, New York
Koolhaas R (2002) Junkspace. October 100:175–190
Kracauer S (1995) The mass ornament: Weimar essays. Harvard University Press, Cambridge
Kuhn TS (1987) A estrutura das Revoluções Científicas. Perspectiva, São Paulo
Kushner D (2012) Machine politics. The New Yorker, 7 May. http://www.newyorker.com/reporting/2012/05/07/120507fa_fact_kushner?currentPage=all. Accessed 8 Nov 2015
Lacan J (1949) Le Stade du miroir comme formateur de la fonction du je, telle que'elle nous est révélée, dans l'expérience psychanalytique. Communication faite au XVIe Congrès international de psychanalyse, à Zurich le 17-07-1949. Rev Fr Psychanal 13(4):449–445
Latham A, Koch R (2013) On the hard work of domesticating a public space. Urban Stud 50(1):6–21
Latour B (1999a) The slight surprise of action: facts, fetishes, factishes. In: Latour B (ed) Pandora's hope: on the reality of science studies. Harvard University Press, Cambridge, pp 266–292
Latour B (1999b) On recalling ANT. Sociol Rev 47(S1):15–25
Latour B (2005) Reassembling the social. Oxford University Press, Oxford
Latour B, Woolgar S (1986) Laboratory life. Princeton University Press, New Jersey
Lazzarato M (1996) Immaterial labour. In: Virno P, Hardy M (eds) Radical thought in Italy. University of Minnesota Press, Minneapolis, pp 133–147
Lazzarato M (2014) Signs and machines. Semiotext(e), Los Angeles
Lefebvre H (1967) Metafilosofia. Civilização Brasileira, Rio de Janeiro
Lefebvre H (1971a) Everyday life in the modern world. Harper & Row, New York
Lefebvre H (1971b) O fim da historia. Dom Quixote, Lisboa
Lefebvre H (1975) Lógica Formal Lógica Dialética. Civilização Brasileira, Rio de Janeiro
Lefebvre H (1976) The survival of capitalism. Allison and Busby, London
Lefebvre H (1991a) Critique of everyday life—volume I. Verso, London
Lefebvre H (1991b) The production of space. Blackwell Publishing, Oxford
Lefebvre H (2000) Right to the city. In: Kofman E, Lebas E (eds) Writings on cities. Blackwell Publishers, Oxford, pp 63–184
Lefebvre H (2014) Toward an architecture of enjoyment. University of Minnesota Press, Minneapolis
Lemley MA (2011) The myth of the sole inventor. Working paper no. 1856610. Stanford Public Law. https://papers.ssrn.com/sol3/papers.cfm?abstract_id=1856610. Accessed 30 Oct 2016
Lenin VI (1961) What is to be done? In: Lenin VI, Lenin's collected works—volume 5. Foreign Languages Publishing House, Moscow, pp 347–530
Lessig L (2001) The future of ideas: the fate of the commons in a connected world. Random House, New York
Lukács G (1971) History and class consciousness. The MIT Press, Cambridge
Lukacs G (2000) A defence of history and class consciousness: tailism and the dialectic. Verso, London
Lukacs G (n.d. [1924]) The Vanguard Party of the Proletariat. https://www.marxists.org/archive/lukacs/works/1924/lenin/ch03.htm. Accessed 2 Jun 2016
Machiavelli N (2008) The prince. Oxford Press, Oxford
Mackay R, Avanessian A (eds) (2014) Accelerate: the accelerationist reader. Urbanomic Publications, Falmouth
Mannheim K (2000) Ideology and utopia. Routledge, Milton Park
Marcuse H (1967) A ideologia na Sociedade Industrial (translation of "One dimensional man"). Ed Zahar, Rio de Janeiro

Marx K(1973 [1857]) Fragments on machines. In: Marx K (1973) Grundrisse. Penguin Books and New Left Review, London
Marx K (1990) Capital—volume 1. Penguin, London
Marx K (1996) Introdução à Crítica da Economia Política. In: Marx K (ed) Marx - Coleção Os Pensadores. Nova Cultural, São Paulo, pp 25–54
Marx K (n.d. [1867]) Letter to Johann Philipp Becker in Geneva, 17 April. https://www.marxists.org/archive/marx/works/1867/letters/index.htm. Accessed 10 May 2013
Marx K, Engels F (n.d. [1848]) Manifesto Comunista. http://www.marxists.org/archive/marx/works/download/pdf/Manifesto.pdf. Accessed 3 Apr 2007
Marx K, Engels F (n.d. [1846]) The German Ideology. https://www.marxists.org/archive/marx/works/1845/german-ideology/ Accessed 10 May 2015
Mazzucato M (2011) The Entrepreneurial State. Demos, London
Minton A (2009) Ground control. Penguin, London
Minton A (2012) Occupy London has lost a battle but started a war over public space. The Guardian, 18 January. https://www.theguardian.com/commentisfree/libertycentral/2012/jan/18/occupy-london-war-public-space. Accessed 18 Dec 2013
Morishima M (1977) Marx's economics: a dual theory of value and growth. Cambridge University Press, London
Negri A (2014) Reflections on the "manifesto for an accelerationist politics". e-flux J 53
Negri A, Hardt M (2000) Empire. Harvard University Press, Cambridge
Németh J (2012) Controlling the commons: how public is public space? Urban Aff Rev XX(X):1–25
Németh J, Schmidt S (2011) The privatization of public space: modelling and measuring publicness. Environ Plan B Plan Des 38:5–23
Newton JD (1987) Uncommon friends: life with Thomas Edison, Henry Ford, Harvey Firestone, Alexis Carrel, & Charles Lindbergh. Harcourt Brace Jovanovich Publishers, New York
Nietzsche F (1987) Sobre o niilismo e o Eterno Retorno. In: Lebrun G (ed) Obras Incompletas - Coleção Os Pensadores. Nova Cultural, São Paulo, pp 425–437
O'Farrell C (2005) Michel Foucault. Sage, London
OMA (2014) Venice Biennale 2014: fundamentals. http://oma.eu/projects/venice-biennale-2014-fundamentals. Accessed 25 Jan 2017
Padgett K (2016) Thomas Edison's Research Laboratory. http://agilewriter.com/Biography/EdisonLab.htm. Accessed 30 Oct 2016
Parker I (2012) High rise—a bold Danish architect charms his way to the top. The New Yorker, September 10. http://www.newyorker.com/magazine/2012/09/10/high-rise. Accessed 12 Sep 2015
Popper K (1989) Conjecturas e Refutações. Ed. UNB, Brasília
Rancière J (1994) The names of history: on the poetics of knowledge. University of Minnesota Press, London
Rancière J (2005) A Partilha do Sensível. Editora 34, Rio de Janeiro
Rancière J (2009) Aesthetic unconscious. Polity, Cambridge
Rancière J (2010) Dissensus. Continuum, London
Read A (2010) Architectural voices. Lund Humphries, London
Rebel Architecture (2014) Al Jazeera documentary series
Rossi A (1982) The architecture of the city. The MIT Press, Cambridge
Rowe C, Koetter F (2000) Collage City. MIT Press, Cambridge
Rynasiewicz R, Renn J (2006) The Galileo affair and the relationship between science and religion. Stud Hist Phil Sci 37(3):259–279
Sallent Del Colombo A (2013) Urban design and the postmodern condition. Urban Stud 50(7):1347–1365
Sande C (2013) Design and the digital commons. Routledge, London
Santos BS (2001) Pela mão de Alice. Cortez, São Paulo
Santos BS (2003) Um Discurso sobre as Ciências. Cortez, São Paulo
Santos BS (2007) Renovar a teoria crítica e reinventar a emancipação social. Boitempo, São Paulo

Schopenhauer A (1909) The world as will and idea. Trübner & Co., London
Schumacher P (2015) The historical pertinence of parametricism and the prospect of a free market urban order. In: Poole M, Shvartzberg M (eds) The politics of parametricism. Bloomsbury, London
Sennett R (2002) Flesh and stone. Penguin Books, London
Shelter (2015a) The housing crisis report. https://www.shelter.org.uk
Shelter (2015b) Housing conditions. http://england.shelter.org.uk/campaigns/why_we_campaign/housing_facts_and_figures/subsection/?section=housing_conditions. Accessed 7 Jul 2015
Shugart S (2001) Performative politics and the social. University of Michigan Press, Ann Arbor
Shvartzberg M (2015) "Play turtle, do it yourself": flocks, swarms, schools, and the political architectural imaginary. In: Poole M (ed) Digital technologies in architecture. Bloomsbury, London, pp 194–222
Siddique H, Abdul G (2025) Yvette Cooper vows to ban Palestine Action under anti-terrorism laws. The Guardian, 23 June. https://www.theguardian.com/world/2025/jun/23/yvette-cooper-vows-ban-palestine-action-anti-terrorism-laws. Accessed 25 Jun 2025
Simmel G (1950 [1903]) The metropolis and mental life. In: Wolff K (ed) The sociology of Georg Simmel. Free Press, New York, pp 409–424
Simondon G (2013) L'individuation à la lumière des notions de forme et d'information. Millon, Grenoble
Simonton DK, Reiter-Palmon R, Tinio P (2015) Thomas Edison's creative career. Psychol Aesthet Creat Arts 9(1):2–14
Singer A (2014) Rebellion in Brazil. New Left Rev 85:19–37
Smith N, Low S (2006) The politics of public space. Routledge, London
Socolovsky M (2012) Fiber bundles, connections, general relativity, and the Einstein-Cartan theory—part II. Adv Appl Clifford Algebras 22(3):873–909
Soiza G (2012) Urban ecology: urban prototype. Urban Ecology Module, MAASD, University of East London [Unpublished student report]
Soja E (2000) Postmetropolis. Blackwell, Oxford
Space Syntax (2011) The social logic of space revisited. http://www.spacesyntax.net
Spencer D (2011) Architectural Deleuzism. Radic Philos 168:9–21
Spencer D (2012) Smooth operators. Unpublished PhD thesis, University of Westminster
Spencer D (2016) Architecture, control, and neoliberal space. Crit Perspect Architect 12(1):45–68
Srnicek N (2013) Technology, capital, value. In: Accelerationism: a symposium on tendencies in capitalism
Staeheli LA, Mitchell D (2008) The people's property? Power, politics, and the public. Routledge, London
Stanek L (2011) Henri Lefebvre on space: architecture, urban research, and the production of theory. University of Minnesota Press, London
Stoetzler M (2016) Intersectional individuality. Sociol Inq 86(2):216–240
Tafuri M (1976) Architecture and utopia. MIT Press, Cambridge
Tafuri M (2006) Interpreting the renaissance. Harvard GSD Publications, Cambridge
Tafuri M, Dal Co F (1976) History of world architecture—modern architecture, vol 1. Electa, New York
Taylor T (2010) The artificial ape: how technology changed the course of human evolution. Palgrave Macmillan, New York
Tedlock B (1991) From participant observation to the observation of participation. J Anthropol Res 47(1):69–94
The Antics Roadshow (2011) Directed by Banksy and D'Cruz J [TV Film]. Channel 4, London
The Competition (2013) Directed by Angel Borrego Cubero [Film]
The Wire (2003) Written by Simon D [TV Series]. HBO, New York
Till J et al (2011) Spatial agency. Routledge, London
Toscano A (2006) The theatre of production. Palgrave, New York

Toscano A (2007) The disparate. In: Forum for European Philosophy annual conference, Society for European Philosophy, University of Sussex, 9 September
Tschumi B (1994) Spaces and events. In: Tschumi B (ed) Architecture and disjunction. MIT Press, Cambridge, pp 139–151
UK Government (2011) Localism Act 2011. https://www.legislation.gov.uk/ukpga/2011/20/enacted. Accessed 12 Mar 2016
UN (2014) Report of the Special Rapporteur on adequate housing as a component of the right to an adequate standard of living—Rolnik R. United Nations General Assembly, Human Rights Council, Twenty-fifth session, Agenda item 3
Vanheule S (2011) Lacan's construction and deconstruction of the double-mirror device. Front Psychol 2(209):1–9
Varna G, Tiesdell S (2010) Assessing the publicness of public space. J Urban Des 15(4):575–598
Virno P (2001) General intellect. In: Zanini A, Fadini U (eds) Lessico Postfordista. Feltrinelli, Milan
Vygotsky (n.d.-a [1925]) The tragedy of Hamlet, Prince of Denmark. Accessed 2 Dec 2016
Vygotsky (n.d.-b [1930]) Mind, consciousness, the unconscious. https://www.marxists.org/archive/vygotsky/works/1930/mind.htm. Accessed 10 Oct 2016
Wainwright O (2014) The truth about property developers. The Guardian, 17 September. https://www.theguardian.com/cities/2014/sep/17/truth-property-developers-builders-exploit-planning-cities. Accessed 14 Jan 2025
Wark M (2004) A hacker manifesto. Harvard University Press, London
Wark M (2013) On telesthesia. Lecture for the center for 21th century studies. University of Wisconsin, Milwaukee
Webb P, Lynch J (2010) Utopian punk. Utop Stud 21(2):313–318
Williams R (1973) Base and superstructure in Marxist cultural theory. New Left Rev 82:3–16
Williams A (2013a) The politics of anticipation. In: Accelerationism: a symposium on tendencies in capitalism
Williams A (2013b) Escape velocities. e-flux J 46
Wilmore J (2022) 'A dangerous gimmick': sector reacts to Right to Buy extension plans. Inside Housing, 9 June. https://www.insidehousing.co.uk/insight/a-dangerous-gimmick-sector-reacts-to-right-to-buy-extension-plans-76019. Accessed 25 Jun 2025
Wolfe R (2012) Memories of future. http://thecharnelhouse.org/2012/08/10/memories-of-the-future/. Accessed 13 Nov 2013
Wood J (2007) Design for micro-utopias. Gower Publishing Limited, Hampshire
Woods M (1999) From craft to profession. University of California Press, Los Angeles
Yoong Z (2014) Fun by design: urban ecology prototype. Urban Ecology Module, MAASD, University of East London [Unpublished student report]
Žižek S (1997) The plague of fantasies. Verso, London
Žižek S (2008) Violence. Picador, New York
Žižek S (2014) Event. Penguin Books, London
Zukin S (2010) Naked city. Oxford University Press, Oxford